William Garnett

Treatise on Elementary Dynamics

for the use of Colleges and schools

William Garnett

Treatise on Elementary Dynamics
for the use of Colleges and schools

ISBN/EAN: 9783337276614

Printed in Europe, USA, Canada, Australia, Japan

Cover: Foto ©berggeist007 / pixelio.de

More available books at **www.hansebooks.com**

A TREATISE

ON

ELEMENTARY DYNAMICS

FOR THE USE OF

Colleges and Schools.

BY

WILLIAM GARNETT, M.A.,

HONORARY D.C.L. OF THE UNIVERSITY OF DURHAM;
(LATE WHITWORTH SCHOLAR;)
LATE FELLOW AND LECTURER OF ST. JOHN'S COLLEGE, CAMBRIDGE;
PRINCIPAL AND PROFESSOR OF PURE AND APPLIED MATHEMATICS IN THE
DURHAM COLLEGE OF SCIENCE, NEWCASTLE-UPON-TYNE.

FIFTH EDITION, REVISED.

CAMBRIDGE:
DEIGHTON, BELL AND CO.
LONDON: GEORGE BELL AND SONS.
1889.

CONTENTS.

CHAPTER I.

ON THE GEOMETRY OF A MOVING POINT, AND THE FUNDAMENTAL LAWS AND PRINCIPLES OF DYNAMICS.

	PAGE
Time	1
Velocity	2
Unit of Length	6
Acceleration	7
Composition of Velocities	9
Relative Velocity	12
Parallelogram of Accelerations	15
Relative Acceleration	17
Angular Velocity	20
Areal Velocity	21
Matter	21
Mass	22
Density	23
Kinetic Energy defined	24
First Law of Motion	25
Second Law of Motion	28
Unit of Force	29
Dyne	30
Impulse	31
Independence of Forces	34
Parallelogram of Forces	35
Weight	36
Value of "g"	36
Weighing	37
Work	38
Horse-Power and Energy	41
Third Law of Motion	43
Conservation of Energy	48
Examples on the Laws of Motion	51
Attwood's Machine	54
Weight Independent of Velocity	57
Weight Proportional to Mass	59
Empirical Formula for "g"	60

vi CONTENTS.

	PAGE
Evidence for Laws of Motion	62
Fundamental Units	63
Change of Units	63
Examples of Change of Units	68
Unit of Length	78
Unit of Area	78
Units of Volume and Time	80
Units of Mass and Velocity	81
Unit of Acceleration	82
Units of Momentum and Force	83
Unit of Work	84
Units of Energy and Density	85
Unit of Impulse	86
Astronomical Unit of Mass	87
British Association Units and Nomenclature	88
Transmission of Power by Belts	91
Transmission of Power by Shafting	93
Examples of Change of Units	94
Examination	97
Examples	100

CHAPTER II.

On Uniform, and Uniformly Accelerated, Motion.

Uniform Velocity	107
Uniformly Accelerated Motion	107
Space described with Uniform Acceleration	108
Geometrical Representation	111
Space described during a Given Interval	114
Equations for Uniformly Accelerated Motion	115
Examples of Uniformly Accelerated Motion	117
Particle sliding down a Rough Plane	119
Pressure of Shot on Target	122
Motion of Particles connected by a String	123
Time of describing a Given Path	132
Time down Chords of a Vertical Circle	133
Straight Line of Quickest Descent	134
Kinetic Energy equivalent to Work done	136
Examples of Kinetic Energy	137
Examples of Horse-Power	139
Convertibility of Energy	140
Conservation of Energy and Perpetual Motion	140
Examples on Units	144
Examination	145
Examples	147

CHAPTER III.

On Projectiles.

Motion of a Projectile	155
Path a Parabola	156

CONTENTS. vii

	PAGE
Range	158
Latus Rectum and Position of Focus	159
Height of Directrix	159
Velocity due to Fall from Directrix	162
Examples of Projectiles	163
Range on Inclined Plane	165
Geometrical Solution of Problems	167
Envelope of System of Trajectories	171
Fountain Jet	171
Change of Kinetic Energy and Work done	173
Problems on Projectiles	174
Examination	180
Examples	181

CHAPTER IV.

ON COLLISION.

Nature of an Impulse	188
Elasticity	189
Coefficient of Elasticity	189
Impact on a Moving Plane	191
Oblique Impact on a Fixed Plane	192
Oblique Impact on a Rough Plane	193
Direct Impact of Smooth Spheres	194
Oblique Impact of Smooth Spheres	197
Action between Elastic Balls	199
"Forces of Compression and Restitution"	200
Examples on Collision	201
Continuous Impact	211
Falling Chain	212
Falling Chain of Variable Density	214
Kinetic Energy after Impact	218
Nature of Impact	221
True Ratio of Forces of Compression and Restitution	223
Ratio of Impulses	223
Energy dissipated by Sudden Changes of Motion	224
Dissipation of Energy in Inelastic Strings	225
Effect of Shocks in Machinery	227
Examination	228
Examples	229

CHAPTER V.

ON CURVILINEAR MOTION, CYCLOIDAL MOTION, THE PENDULUM, NORMA ACCELERATION, CENTRIFUGAL FORCE, INITIAL TENSIONS, ETC.

Motion of a Particle in a Smooth Tube	243
Newton's Experiments on Impact	246
Properties of the Cycloid	249
Motion of a Particle on a Cycloid	251
Length of Arc of a Cycloid	253

CONTENTS.

	PAGE
Cycloidal Pendulum	254
"Simple Equivalent Pendulum"	256
Determination of "g" by Pendulum	257
Normal Acceleration	258
Centrifugal Force	261
The Conical Pendulum	261
The Parabolic Governor	264
Tension in a Running Belt	266
Velocity of a Transverse Wave in a Stretched String	267
Harmonic Motion	268
Motion of certain Systems of Bodies	270
Initial Actions	273
Pressure of a Steam Hammer	277
Examination	279
Examples	280

CHAPTER VI.

ON THE DYNAMICAL THEORY OF GASES.

Constitution of a Simple Gas	286
Pressure due to a Stream of Particles	288
Mean value of $\cos^2 a$	289
Velocity of Mean Square	293
Boyle's Law	295
Gay Lussac's Law	295
Velocity of Mean Square of Hydrogen	296
Absolute Zero of Temperature	297
Work done by Gas in Expanding	297
Cooling produced by Expansion	299
Expansion into Vacuum (Joule's Experiment)	300
Graham's Law of Diffusion	301
Gases act to one another as Vacua	301
On the Average Value of $\cos^2 a$	302
Miscellaneous Examples	303

ANSWERS 313

ELEMENTARY DYNAMICS.

CHAPTER I.

ON THE GEOMETRY OF A MOVING POINT, AND THE FUNDAMENTAL LAWS AND PRINCIPLES OF DYNAMICS.

1. A POINT is said to be in motion when it changes its position relative to surrounding objects.

From this definition it will be seen that all cases of motion which will come under our consideration are essentially *relative*; in fact, we have no means of measuring *absolute* motion, or of determining whether any given point is absolutely at rest in space, or not, and it may even be doubted whether the human mind is capable of forming a distinct conception of *absolute* motion.

We shall not attempt to give a definition of *space*; our idea of it must be considered a primary conception.

Time is defined by the metaphysician as "the succession of ideas"; the physicist treats time, like space, as a primary conception.

Equal times are generally defined as those intervals during which the earth turns through equal angles relative to the fixed stars, and any duration of time may then be measured by the angle turned through by the earth during the interval. The most obvious unit of time is therefore the sidereal day, or the period during which the earth makes a complete rotation on its axis relative to the fixed stars. The unit generally adopted is the second of mean solar time.

That our fundamental conception of the measurement of time is not, however, based upon the rotation of the earth is apparent from the fact that we sometimes ask whether the length of the day has changed during the last two or three thousand years. Were the definition of equal times to which we have just alluded generally accepted, this question would be absurd, since all days would be equal by definition. The test of equality between two intervals of time will be discussed in connection with the first law of motion, meanwhile the definition above given will be found sufficient for our purpose.

2. The velocity of a point is the *rate* at which it is changing its position relative to surrounding objects; in other words, the degree of speed with which it is moving.

A point is moving with *uniform* velocity when it passes over equal distances in equal intervals of time: under other circumstances its velocity is said to be *variable*.

A distinction has been drawn between *velocity* and *speed* by restricting the use of the word speed to cases in which the direction of the motion is not considered while velocity implies direction as well as degree of quickness. If this distinction were generally adopted we could not speak of a body revolving in a circle with uniform velocity but only with uniform speed, while the velocity would be undergoing a constant change. Uniform velocity would imply uniformity in direction as well as in magnitude.

3. If we wish to convey the idea of *speed* in speaking of the motion of anything we say that it passed over a certain distance (say 50 miles) in a certain time (say an hour); while, if we wish to convey an idea of the slowness of the motion of anything we say that it took a certain time (say an hour) to traverse a certain distance (say two miles). Now the velocity of a point, being defined as the degree of *speed* with which it is moving, must always be expressed according to the former method; *viz.* as so many units of length per unit of time.

4. The complete representation of any physical quantity consists of two factors, one of which is the unit in

terms of which the quantity is measured and must be of the same kind as the quantity itself, while the other is a pure number indicating the ratio of the quantity to this unit and called its *measure*. Velocity like all other quantities must be measured by its ratio to a unit of its own kind, that is, to a certain velocity selected as the standard, and the magnitude of this unit is all that is in our power to select. The velocity of a point which, moving uniformly, passes over the unit of length in the unit of time is taken as the unit of velocity, and the velocity of any other point is measured by its ratio to this unit. In this case the measure of the velocity of a point which is moving uniformly will be equal to the number of units of length traversed by it in the unit of time.

5. In order to measure the velocity of a point which is moving uniformly we have but to measure the space traversed in a given time, and dividing this distance by the measure of the time we have the space traversed during each second, or in other words, the velocity of the point.

6. When the velocity of a point changes continuously we may speak of its value at some particular instant, but a little care is required in order to determine exactly what is meant by the measure of a variable velocity at any particular instant. If the velocity be increasing, then, during the second succeeding the instant in question the point will move over a greater space than if the velocity were not on the increase but remained the same as at the commencement of the second. On the other hand, during the preceding second the space traversed is less than if the velocity were constant throughout the second and the same as at the end of it. We shall obtain a better result if we observe the space traversed during the second which contains the instant in question and half of which precedes while the other half succeeds the instant, but this will not give an accurate result unless the velocity changes uniformly. If however we take a very short interval including the proposed instant, the mean or average velocity during that interval will be obtained by dividing the measure of the distance traversed

by that of the interval. Now if the interval taken be exceedingly short the velocity has no opportunity of changing sensibly during it, and the mean velocity during the interval cannot sensibly differ from the velocity at any instant contained therein. Hence, by making the interval sufficiently short, we can obtain a result which differs by as little as we please from the velocity at the proposed instant. If we could make the interval indefinitely short and still perform the above operation we should realize our conception of the velocity of the point at a particular instant.

7. It is usual to state that the velocity of a point at a particular instant is measured when variable by the space which would be passed over in the unit of time supposing the velocity constant during the unit and *the same as at the proposed instant*. The words in italics take us back to the original difficulty, so that we appear to gain very little by the definition.

8. If the unit of length be increased or decreased, the unit of time remaining the same, the space passed over in the unit of time by a point moving with unit velocity is increased or decreased, and therefore the unit of velocity is changed, in that same ratio. If, on the other hand, the unit of time be increased or decreased, the unit of length remaining the same, the time required by a point moving with unit velocity to pass over the unit of length is increased or decreased accordingly. Now the longer the time occupied by a point in moving over the same distance, the less must be its velocity, and the shorter the time the greater the velocity. Hence the unit of velocity must vary inversely as the unit of time, if the unit of length remain constant. Also we have just shown that the unit of velocity varies directly as the unit of length when the unit of time is kept constant. Therefore, when all are allowed to vary together, the unit of velocity will vary directly as the unit of length, and inversely as the unit of time. (See Todhunter's *Algebra*, Art. 425.)

9. As above stated the mathematical expression for any physical quantity always consists of two factors, one

being the unit of the same kind as the thing considered, the other representing the number of such units in the quantity considered, and constituting the numerical measure of such quantity. The complete representation of a physical quantity in mathematical language must therefore consist of two symbols, representing these two factors respectively. The unit is sometimes represented by a capital letter placed in square brackets; *e.g.* the unit of time thus [T]. Now the equations used in almost all mathematical investigations are equations between the *numerical measures* of quantities, and *not* between the *quantities* themselves; the symbol representing the unit is therefore omitted, since the unit itself does not enter into the equations, and in consequence, the habit of representing only the numerical measures of quantities has become so general, that even when the quantities themselves are considered, but one symbol is generally used, the corresponding unit being understood.

Now any quantity being represented by the product of the unit of the same kind, and the number of such units contained in the quantity considered, it is obvious that if the unit change, the quantity measured remaining the same, the number of units contained in it will be changed in the inverse ratio of the unit. Hence the numerical measure of any given quantity varies inversely as the unit in terms of which it is measured. For example, a stick whose length is 72 *inches* will measure $\frac{72}{12}$, or 6, *feet*, and $\frac{72}{36}$, or 2, *yards*.

10. Applying the principles of the preceding article to the measurement of velocity, we see that the numerical measure of any velocity will vary inversely as the unit of velocity; and it has been shown that the unit of velocity varies directly as the unit of length and inversely as the unit of time. Hence the numerical measure of a velocity varies inversely as the unit of length and directly as the unit of time. Thus, a velocity of 10,560 feet per minute is equivalent to a velocity of 2 miles per minute, or of 120 miles per hour, or of 176 feet per second. Other examples

of the change of units will be found at the end of this chapter.

The British standard unit of length is the imperial yard, which is defined to be the distance between the centres of the lines engraved on two gold plugs in a bronze bar kept in the Exchequer chambers, and known as the imperial standard yard, the temperature of the bar being 62° Fahrenheit. The unit generally adopted by engineers is one-third of this distance, and is called a foot. As before stated, the unit of time generally adopted is the second of mean solar time.

11. If distances measured in one direction along a line be considered positive, it is usual to consider distances measured in the opposite direction as negative. Thus if distances measured from A towards B be reckoned posi-

$$\overline{\qquad C \qquad\qquad A \qquad\qquad\qquad B \qquad}$$

tive, those measured from B towards A or from A towards C will be negative. The same convention is extended to velocities. If a point move along a line in the direction in which distances are reckoned positive, its velocity is considered positive, but if it move in the opposite direction its velocity is considered negative. Thus, if a point move from A to B, it increases its distance from A, measured in the positive direction, and its velocity is accordingly positive; but if it move from B towards A it diminishes its distance from A, and its velocity is considered negative. Again, if the point move from A towards C, though it increases its distance from A considered *numerically*, yet such distance being negative is decreased *algebraically*, and the velocity of the point is accordingly reckoned negative. In a similar way it will be seen that the velocity of a point moving from C towards A will be positive. This convention being adopted, the velocity of a point which continues to move in the same direction will not change sign when the point passes through A. This is, of course, as it should be, since there is no more reason why the velocity of a point should change sign when the point passes through A than at any other point of its path.

UNIT OF ACCELERATION.

12. *Acceleration* is the rate of change of velocity.

It is said to be *uniform* when equal increments of velocity are generated in equal intervals of time. If this be not the case the acceleration is *variable*.

It is measured, when uniform, by the velocity generated in a unit of time; when variable, it is measured, at any instant, "by the velocity which would be generated in a unit of time, were the acceleration to remain constant during that unit, and the same as at the proposed instant." All that has been said respecting variable velocity applies to the measurement of variable accelerations.

In order that the above may furnish a proper measure of acceleration, it will be seen that the unit of acceleration must be that of a point whose velocity is increased by the unit of velocity in the unit of time. If the unit of length vary, the unit of time remaining the same, it has been shown that the unit of velocity will vary in the same ratio; hence, the unit of acceleration will also vary in the same ratio, and therefore when the unit of time remains constant, the unit of acceleration varies directly as the unit of length.

13. Next suppose the unit of length to remain constant, but the unit of time to vary. Then we have seen that the unit of velocity varies inversely as the unit of time. Now if the acceleration were always allowed the same time for the generation of the unit of velocity, the unit of acceleration would then vary directly as the unit of velocity, that is, inversely as the unit of time. But the time allowed for the generation of the unit of velocity does not remain constant; it is in fact the *unit* of time, and therefore varies as that unit varies. Now if the time be diminished in which any *given* velocity is generated the acceleration must be proportionately increased; and if the time be increased, the acceleration must be proportionately diminished. Hence, if the unit of velocity could be kept constant, the unit of acceleration would vary inversely as the unit of time, simply because the time during which the unit of velocity must be generated is changed. But it has been shown that if this latter were kept constant, the unit of acceleration would vary inversely as the unit of time, solely on account of the change in the unit of velocity. Hence, taking both reasons into account, when the unit of length remains constant, the unit of acceleration must vary inversely as the square of the unit of time. Also it has been shown that the unit of acceleration varies directly as the unit of length when the unit of time remains constant. Therefore, when all three are allowed to vary together, the unit of acceleration must vary directly as the unit of length, and inversely as the square of the unit of time.

14. The numerical measure of any acceleration varies inversely as the unit of acceleration. Therefore the numerical measure of a given acceleration varies inversely as the unit of length and directly as the square of the unit of time.

The above reasoning will be rendered much clearer by the consideration of an example.

Suppose a certain acceleration to be represented by 32 when a second and a foot are the units of time and length respectively; what will be the measure of the same acceleration when a minute and a yard are units?

With the acceleration

in $1''$ there is generated a velocity per $1''$ of 32 feet;
∴ ,, $1''$,, ,, ,, ,, $60''$,, 32×60 feet;
∴ ,, $60''$,, ,, ,, ,, $60''$,, 32×60^2 feet.

But 32×60^2 feet are equivalent to 38,400 yards. Hence with the given acceleration in one minute there is generated a velocity of 38,400 yards per minute, and therefore when a minute and a yard are the units of time and length respectively the acceleration will be represented numerically by 38,400.

15. An acceleration is reckoned positive when the velocity of the moving point tends to increase algebraically; if the velocity tend to diminish algebraically the acceleration is considered negative. Thus an acceleration is considered positive if, with it, a positive velocity increase numerically or a negative velocity decrease numerically; while an acceleration with which a positive velocity decreases or a negative velocity increases numerically is

considered negative. An acceleration with which the *numerical* measure of a velocity tends to decrease is frequently called a retardation.

16. A velocity is completely known if we know its magnitude and direction. Now a straight line can be drawn in any direction and of any length; if then a straight line be drawn in the direction in which a point is moving and of such length as to contain as many units of length as there are units of velocity in the velocity of the point, such a straight line will represent in every respect this velocity.

Similarly an acceleration is completely known if we know its magnitude and the direction of the velocity generated. An acceleration may therefore be represented in every respect by a straight line drawn in the direction of the velocity generated, and containing as many units of length as there are units of acceleration in the acceleration in question. Moreover, since an acceleration is measured by the number of units of velocity generated in the unit of time, the straight line which represents an acceleration in magnitude and direction may also completely represent the velocity generated in the unit of time to which the acceleration corresponds.

17. Suppose a point to be moving with two independent velocities in any directions; then at any instant the point must be moving in some definite direction and with some definite velocity; this velocity must therefore be equivalent to the two independent velocities, and is called their resultant; the independent velocities themselves, considered with reference to their resultant, are called components. The same reasoning must apply to any number of independent velocities with which a point may be moving, and which must therefore be equivalent to a single resultant. Let a point be supposed to move with *two* independent velocities, and let it be required to find the actual or resultant velocity of the point. For example, the point may move with a given velocity along a straight tube, while the tube, always remaining parallel to its original direction, slides with a given velocity and in a given direction along a fixed plane. The problem

will then be to determine the velocity of the point and the direction of its motion relative to the fixed plane. The two independent velocities in this case are the velocity of the point along the tube, and the velocity of the tube relative to the fixed plane.

18. If the two velocities be in the same straight line it is obvious that the resultant velocity is the algebraical sum of the two; velocities in one direction being considered positive and those in the opposite direction negative; and similarly the resultant of any number of independent velocities in the same straight line is the algebraical sum of the component velocities.

If the independent velocities be not in the same straight line their resultant must be found by help of the following proposition, known as the "parallelogram of velocities."

PROP. *If a point be moving with two independent velocities represented in magnitude and direction by two straight lines drawn from a point, the resultant velocity will be represented in magnitude and direction by the diagonal, drawn from that point, of the parallelogram constructed upon these two straight lines as adjacent sides.*

Let AB, AC represent in magnitude and direction the velocities, which we denote by u, v respectively. Then

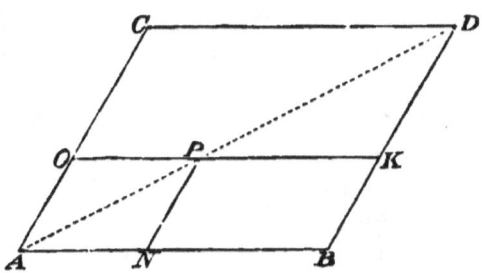

AB, AC may denote the spaces passed over in the unit of time by points moving with the velocities u, v respectively. Complete the parallelogram $ABDC$, and draw the diagonal AD. Then AD shall represent in magnitude and direction the resultant of the velocities represented by AB and AC respectively.

Let the moving point be denoted by P, and suppose P to move along a straight tube OK with uniform velocity u, while the end O of the tube moves uniformly along the straight line AC, with velocity v, the tube remaining always parallel to AB. Then, supposing the point P to start from A when the tube is in the position AB, if at the end of any time τ we take, along AC, AO equal to $v\tau$, and draw OK parallel to AB, OK will be the position of the tube; and taking OP equal to $u\tau$, P will be the position of the point. Draw PN parallel to AC; then

$$AN = OP = u\tau;$$
$$\therefore AN : AO :: u\tau : v\tau :: u : v$$
$$:: AB : AC.$$

Therefore the parallelograms ON, CB are similar. But similar parallelograms which have a common angle are about the same diagonal; therefore P lies on the diagonal AD; and since τ may be any interval we please, the point P will always lie on AD or AD produced; in other words, AD is the path of P, and therefore always represents its resultant velocity in direction. Also the resultant velocity of P is constant in magnitude: for AP, the space passed over during the interval τ, is always proportional to AN or $u\tau$, that is to τ, since u is constant. The space passed over in any time is therefore proportional to that time, and the velocity of the moving point is therefore constant in magnitude.

Again, at the end of the unit of time after leaving A the end O of the tube will have reached C, and the point P will consequently be at D. Hence in one unit of time P will have moved from A to D, and we have just shown that its path is the diagonal AD and its velocity uniform. Hence the straight line AD represents in magnitude and direction the resultant velocity of the point P, that is, the resultant of the velocities represented by AB and AC respectively.

Therefore, if, etc. Q.E.D.

19. If a particle be moving with more than two independent velocities, we can find their resultant by finding first the resultant of any two, then compounding that with a third, and so on. Also, since velocities, like forces, are subject to the parallelogram law of composition and resolution, the propositions which are true for a system of forces acting at a point *on account of the forces being subject to this law*, are also true for a number of independent velocities with which a point may be moving. Thus, if the straight lines representing the velocities are equal, and parallel to, and in the same sense as, the sides of a closed polygon taken in order, the point is at rest; if they form all but one of the sides taken in order of such a polygon, their resultant is represented by the remaining side taken in the reverse order. Similarly we have the triangle of velocities, the parallelepiped of velocities, and so on.

20. If the velocities of two moving points, A and B, be given relative to certain points which we consider fixed, we can, by help of the "parallelogram of velocities," determine the velocity of A relative to B.

Suppose, for example, that B is a point fixed on some surface which moves in any direction without rotation, and with known velocity, so that every point of the surface has the same velocity as B, and that A is a point which moves on that surface, the velocity of A, at any instant, relative to fixed objects around being known; then the velocity of A relative to the surface, that is, relative to B, can be found. For instance, B may be a point fixed on the deck of a ship which is moving uniformly on a still sea, and A some point moving about on deck; the velocities of A and B relative to the water being given, the velocity of A relative to the ship's deck can be found.

If the points A and B be moving in the same straight line and in the same direction with uniform velocities u and v respectively, it is obvious that the distance between A and B is increased or diminished during each unit of time by a space numerically represented by $u-v$, according as A is in front of, or behind, B; $u-v$ is therefore the velocity of A relative to B. If $u-v$ be negative, it shows that the velocity of A relative to B is in the direction opposite to that in which the points are moving.

If A and B be moving in *opposite* directions with velocities represented *numerically* by u and v respectively, it may be shown that the velocity of A relative to B is *numerically* represented by $u+v$; but if u and v represent not only the numerical but the algebraical values of these velocities, then u and v will be of opposite signs, and the velocity of A relative to B will be represented by $u-v$, as in the case in which A and B are moving in the same direction.

21. In each of the above cases the velocity of A relative to B may be found by the following process. Let a velocity equal and opposite to that of B be supposed given to both A and B. Then B will be brought to rest, and the *same* velocity being impressed on both A and B, their *relative* velocity will be unaffected. For example, if a man be walking on the deck of a ship, his velocity relative to the deck, or any point upon it, is altogether independent of the ship's motion, and will remain the same if the ship be brought to rest. The velocity of A will then be $u-v$; and since B will have been brought to rest, this will be the velocity of A relative to B. Hence if two points A and B be moving in the same straight line with velocities represented in magnitude and direction by u and v respectively, the velocity of A relative to B will be represented in magnitude and direction by $u-v$.

If A and B be moving in parallel straight lines with velocities represented respectively by u and v, we may show, by precisely the same method as that adopted above, that the velocity of A relative to B is $u-v$, as before.

22. Suppose A and B to move with velocities repre-

sented by u and v respectively, but not in the same or parallel straight lines. Let the velocity of A be represented in magnitude and direction by OH, and that of B by OK. Then the velocity of A relative to B will be represented by KH.

For the velocity of A may be considered as the resultant of two velocities; *viz.* the velocity of B, and the velocity of A relative to B. Hence the velocity represented by OH is the resultant of two independent velocities, one of which, that of B, is represented by OK, and the other is the velocity of A relative to B. But the velocity represented by OH is, by the "parallelogram of velocities," the resultant of velocities represented by OK and KH. Therefore the velocity of A relative to B is represented in magnitude and direction by KH.

If this velocity be denoted by w, we have, since
$$KH^2 = OH^2 + OK^2 - 2OH \cdot OK \cos KOH,$$
$$w^2 = u^2 + v^2 - 2uv \cos HOK.$$

23. The same result may be obtained in a different way. As before, let OH, OK represent the velocities of

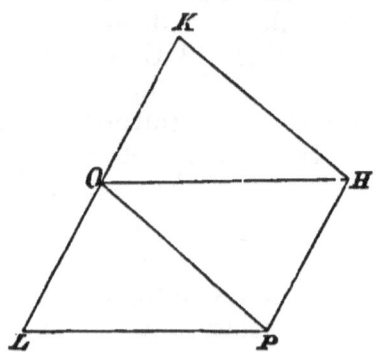

A and B respectively. Suppose a velocity equal and opposite to that of B to be impressed upon both A and B. Let OL represent in magnitude and direction this velocity. Then OL is equal to KO, and in the same straight line with it. Also by this means B will be brought to rest while A is made to move with two independent velocities; *viz.* its original velocity represented by OH and the velocity we have supposed impressed upon it, which is represented by *OL*. Complete the parallelogram *OHPL*; then, by the "parallelogram of velocities," the diagonal *OP* represents in magnitude and direction the resultant velocity of *A*. But since *B* is now at rest, the resultant velocity of *A* is the same as its velocity relative to *B*. Also the velocity of *A* relative to *B* will have been unchanged by impressing the same velocity on both *A* and *B*. Hence *OP* represents in magnitude and direction the velocity of *A* relative to *B*.

If, then, we wish to find the velocity of a moving point *A* relative to another moving point *B*, we may impress on both *A* and *B* a velocity equal and opposite to that of *B*, and the resultant velocity of *A* will then be the velocity required.

That this process leads to the same result as the method of the preceding article is at once obvious, for *KOPH* is a parallelogram, and therefore *OP* is equal and parallel to *KH*.

If the velocities of *A* and *B* be not uniform, the velocity of *A* relative to *B* at any *instant* may be found as above, *OH* and *OK* representing in this case the velocities of *A* and *B* respectively *at the instant* in question.

The velocity of *B* relative to *A* is of course equal and opposite to that of *A* relative to *B*.

24. Accelerations, also, like forces and velocities, may be resolved and compounded according to the parallelogram law. This we proceed to prove.

PROP. *If a point be moving with two independent accelerations represented in magnitude and direction by two straight lines drawn from a point, the resultant acceleration will be represented in magnitude and direction by the diagonal, drawn from that point, of the parallelogram constructed on the two straight lines, representing the accelerations, as adjacent sides.*

For simplicity of expression, suppose a second to be the unit of time. Let *AB* represent the initial velocity of the point; *BC*, *BD* the accelerations. Then *BC*, *BD* represent the velocities generated in one second corresponding to the respective accelerations taken singly. Now since the accelerations are independent, their combined effect produced at any instant is the sum of the effects corresponding to each considered singly at that instant, and

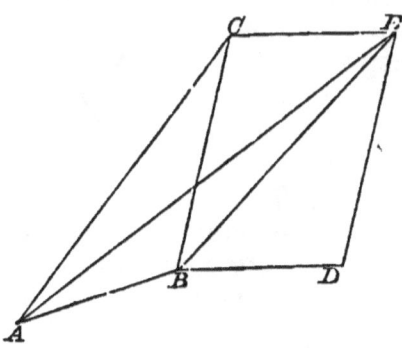

hence the final change of velocity produced in one second under the two accelerations together is the same as if each had existed separately during one second, since the effect of an acceleration is independent of the velocity of the moving point. Suppose then the acceleration represented by BC to exist by itself for one second; the velocity generated in that interval is represented by BC, and compounding this with the original velocity, AB, of the moving point, it follows from the parallelogram of velocities that the resultant velocity is represented by AC. Now suppose the acceleration represented by BD to exist singly for a second, the point moving, initially, with the velocity represented by AC. Draw CE equal and parallel to BD; then CE represents the velocity generated in one second corresponding to the acceleration represented by BD. Combining this with the velocity which the point already possesses, viz. that represented by AC, the final velocity is, by the parallelogram of velocities, represented by AE. But the effect when the two accelerations exist together for one second is the same as the whole effect produced when each exists separately for one second. Hence the effect when the two accelerations exist together for a second is to change the velocity of the moving point from that represented by AB to that represented by AE. The velocity generated when the two accelerations exist together for a second is therefore represented by *BE*, since *BE* represents that velocity which, when compounded with the velocity represented by *AB*, produces a resultant represented by *AE*. But the whole velocity generated in a second is the measure of the resultant acceleration; hence *BE* represents the acceleration which is the resultant of the accelerations represented by *BC* and *BD*. But *BE* is the diagonal of the parallelogram constructed upon the straight lines *BC*, *BD* as adjacent sides. Therefore, if, etc. Q.E.D.

Hence, accelerations, like forces and velocities, are subject to the parallelogram law of composition and resolution, and any propositions true of forces *in consequence of their being subject to this law* must also be true of accelerations. We have therefore the triangle, polygon, etc., of accelerations, and any number of accelerations may be compounded in the same way as a system of forces acting at a point.

25. By help of the preceding proposition, if the accelerations of moving points *A* and *B* be given, we can, as in the corresponding case of velocities, find the acceleration of *A* relative to *B*.

Let *OH*, *OK* represent in magnitude and direction the accelerations of *A* and *B* respectively. Suppose an acce-

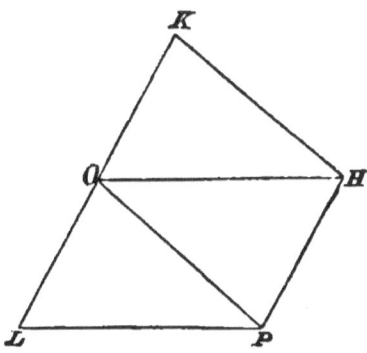

leration equal and opposite to that of *B* impressed upon both *A* and *B*. This cannot affect the relative motions of *A* and *B*, and therefore does not affect their relative accelerations. (For we may suppose A and B to be points moving about inside a closed surface; then if the same motion in space be impressed upon the surface and everything within it, this will obviously not affect the motions of A and B relative to the surface or each other.) Let OL represent the acceleration equal and opposite to that of B. Then OL is equal to KO and in the same straight line with it. Now an acceleration equal and opposite to that of B having been impressed upon it, it will be moving with no acceleration, that is, with uniform velocity. Suppose a velocity equal and opposite to that of B impressed on both A and B; this will not affect their relative velocity, but B will thereby be brought to rest. Complete the parallelogram $OHPL$, and draw the diagonal OP. Then the acceleration of A is the resultant of two independent accelerations represented respectively by OH and OL, and is therefore, by the parallelogram of accelerations, represented in magnitude and direction by the diagonal OP of the parallelogram constructed upon OH and OL; and, since B is now at rest and not possessed of any acceleration, this is the acceleration of A relative to B.

Since $KOPH$ is a parallelogram, OP is equal and parallel to KH; hence KH represents in magnitude and direction the acceleration of A relative to B.

From the above investigation we see that in order to find the acceleration of a moving point A, relative to another moving point B, we have only to impress on both A and B an acceleration equal and opposite to that of B. The resultant acceleration of A is then the original acceleration of A relative to B. Or, more briefly, if the accelerations of A and B be represented in magnitude and direction by two straight lines OH, OK respectively drawn from a point O, the acceleration of A relative to B will be represented in magnitude and direction by the straight line KH.

26. If the accelerations of A and B be not uniform, the acceleration of A relative to B may be found *at any instant* by the above process, OH, OK representing the accelerations of A and B respectively *at that instant*.

The acceleration of B relative to A is obviously equal and opposite to that of A relative to B.

If the accelerations of A and B be in the same straight line and be represented *numerically* by a and β respectively, the acceleration of A relative to B will be represented by $a \pm \beta$, according as the accelerations of A and B are in the same or in opposite directions. If the accelerations be represented *algebraically* by a and β, the acceleration of A relative to B will be represented in magnitude and direction by $a - \beta$.

27. If a point move with uniform velocity v, the space passed over by it in t units of time is equal to vt units of length. For the distance passed over in each unit of time is v units of length, and therefore the distance passed over in t units of time is vt units of length.

If a point move with a certain acceleration always in the direction of motion and represented by f, the velocity generated in t units of time and corresponding to this acceleration is represented by ft. For the velocity generated in each unit of time is numerically equal to the acceleration, and is therefore f units of velocity; therefore the velocity generated in t units of time is ft units of velocity.

If the direction of the acceleration be always that of the motion of the point, and its magnitude be constant and represented by f, and if the point be originally moving with u units of velocity, its velocity after t units of time will be represented by $u + ft$.

If a point be moving with an initial velocity in a certain direction, and then continue moving for a given time with an acceleration in some other constant direction, since with a given acceleration the velocity generated at any instant is independent of the velocity or direction of motion of the moving point, it follows that the final velocity of the point may be determined by finding the velocity generated in the direction of the acceleration, and compounding this with the original velocity of the point according to the parallelogram law.

28. If P represent a point which moves in any manner

relative to O, and OX be a line through O and fixed in

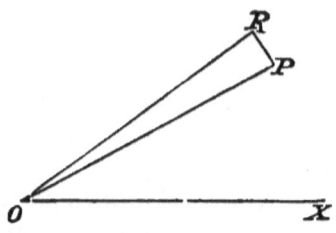

direction, the rate of change of the angle POX is the angular velocity of P about O.

If P be moving directly towards or from O its angular velocity about O is zero.

If P be moving at right angles to OP with velocity v, and P_1 be its position after a very short time τ, then $PP_1 = v\tau$ and the angle POP_1 described in time τ is represented by $\dfrac{PP_1}{OP} = \dfrac{v\tau}{OP}$. Hence the angular velocity of P about O is represented by $\dfrac{v}{OP}$.

29. If P be moving with velocity v in a direction making an angle θ with OP, and P_2 be its position after

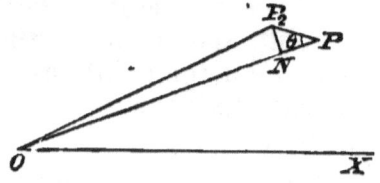

a very short time τ, then PP_2 is represented by $v\tau$. Draw P_2N perpendicular to OP. Then
$$P_2N = PP_2 \sin\theta = v\tau \sin\theta.$$
Also the angle $POP_2 = \dfrac{P_2N}{OP_2} = \dfrac{v\tau \sin\theta}{OP_2}$ and the angular velocity about O is represented by $\dfrac{v \sin\theta}{OP_2}$ or $\dfrac{v \sin\theta}{OP}$, since by taking τ sufficiently small OP_2 may be made to differ from OP by a quantity as small as we please.

KINETICS. 21

30. If the point P is moving along any line XP (which may be straight or curved) the rate of increase of the area POX is called the areal velocity of P about O.

Referring to the figure of the last article, the area of the triangle POP_2 is represented by $\frac{1}{2}P_2N \cdot OP = \frac{1}{2} \cdot v\tau \sin\theta \cdot OP = \frac{1}{2} v\tau \sin\theta \cdot r$ if OP be represented by r. Hence the areal velocity will be represented by $\frac{1}{2} vr \sin\theta$.

If ω represent the angular velocity, then $\omega = \dfrac{r \sin\theta}{r}$ or $v \sin\theta = \omega r$. Hence the areal velocity is represented by $\frac{1}{2}\omega r^2$.

31. The rate of change of angular velocity is called angular acceleration. It is measured when uniform and when variable according to the same principles as linear acceleration.

Rate of change of areal velocity is called areal acceleration.

KINETICS.

32. Hitherto we have been considering simply motion without any reference to the agencies producing it, or the properties of the thing moved. This portion of the subject, or the geometry of motion, is frequently called *Kinematics*. We must now consider motion with reference to the agencies producing it, and the things in which it is produced. The term *Dynamics* is sometimes confined to this section of our subject, but it is more frequently known as *Kinetics*, while *Dynamics* is (improperly) understood to embrace both this and *Kinematics*.

Matter, like space and time, must be considered as one of the inevitable primary conceptions of the mind, of which no satisfactory definition can be given. Many of its properties are known to us with more or less of scientific exactitude by our every-day experience, and it is to this we must refer for a distinct conception of *matter*.

33. The most characteristic manner in which matter affects our senses is through the effort required to produce in it sudden changes of motion. If a small mirror be held

in the hand a reflected sunbeam may be made to dance about in any arbitrary manner without any sensible effort on the part of the operator, for the small exertion of which he is conscious is the same whether the sun be shining on the mirror or not. But it is far from easy to move a half hundred-weight quickly aside and then sharply to bring it to rest, even though it be suspended by a very long string which supports its weight and allows of its being moved almost in a horizontal plane. We say that the half hundred-weight is matter, while we call the sunbeam immaterial.

The character of matter, in virtue of which an effort is required to rapidly change its motion, is sometimes called inertia, but as this term only conveys the same idea as the word *mass*, when properly understood, it will not be frequently employed in subsequent pages.

34. A *body* is a quantity of matter limited in every direction.

A *particle* is a portion of matter whose dimensions in every direction are indefinitely small, and which may therefore be treated as a physical point. We however adopt the convention of conceiving particles to contain a finite quantity of matter, though it is contrary to experience that an indefinitely small body should contain a finite quantity of anything. The conception of a particle simply enables us to treat a small body as though it were indefinitely small, and thus enables us to neglect motions of rotation which, if considered, would remove our investigation to the domain of "Rigid Dynamics."

Force is that which produces or tends to produce motion in matter, or modifies or tends to modify existing motion.

The *mass* of a body or particle is the quantity of matter which it contains.

35. The masses of two or more particles are said to be *equal* when the same force acting similarly upon them for the same time generates in them the same velocity.

If two bodies of equal mass be connected so as to form one, the mass of the body so formed is double that of each of its constituents. Similarly, if three equal masses be connected together, we get a body of triple mass, and so on. We thus arrive at a system of measurement applicable to masses. The British standard unit of mass is the imperial standard pound Avoirdupois, a mass of platinum kept at the Exchequer chambers. All masses may be expressed in terms of this unit, or of other units deduced therefrom. The French imperial standard of mass is Borda's platinum kilogramme, which was originally constructed so as to contain the same amount of matter as a cubic decimetre of distilled water at the temperature corresponding to its maximum density. The pound and the kilogramme are standard of *mass*, not of *force*. After stating the laws of motion we shall see how unequal masses may be compared by observing the effects of forces upon them.

We are not at present in a position to explain any system of measurement applicable to forces. Newton's second law of motion will however provide us with the means of comparing forces, and consequently of measuring them in terms of some unit.

36. The *density* of a body, when uniform, is the ratio of the mass of any volume of the body to that of an equal volume of some standard substance.

When variable, an approximation to the density at any point is obtained by determining the number of units of mass in a very small volume including the point and dividing by the measure of the volume. If the volume could be indefinitely diminished the result would give accurately the density at the point in question. Hence, when the density of a body is variable its measure at any point is the ultimate ratio of the number of units of mass in any volume containing that point to the number of units of volume in the same when such volume is indefinitely diminished.

The standard substance is generally so chosen that the unit of volume of the substance contains the unit of mass of matter. The unity of density is consequently the density of a uniform substance, of which the matter contained in the unit of volume is the unit of mass. Then, the density of a body, whose density is uniform, will be measured by the number of units of mass contained in the unit of volume, or, which is the same thing, the ratio of the number of units of mass contained in any portion of the body to the number expressing the volume of that portion.

If the density of a substance be uniform, and numerically represented by ρ, the mass of V units of volume of the substance will be $V\rho$ units of mass.

For each unit of volume contains ρ units of mass, and therefore V units of volume will contain $V\rho$ units of mass.

37. The *momentum* of a particle is the product of its mass and its velocity. The unit of momentum is consequently the momentum of the unit of mass moving with the unit of velocity, and the momentum of any moving mass is measured in terms of this unit. The phrase "quantity of motion" was used by Newton in place of the more modern term "momentum."

38. The *vis viva* or *kinetic energy* of a moving particle may be defined as one half the product of its momentum and its velocity.

The vis viva used to be defined as the product of the mass and the square of the velocity of the moving particle, and the kinetic energy was called the *semi-vis viva*; but the above definition is more convenient, and has moreover the advantage arising from the fact that both the momentum and the velocity of a particle admit of a physical interpretation, while no meaning has been assigned to the square of a velocity.

39. All the *theorems* hitherto given have been deduced from abstract reasoning, and it is impossible for us to conceive of any order of things in which these theorems should not be true; but in order to determine the mutual relations between force and matter, or, in other words, the effect of forces upon matter, we must have recourse to experiment. The conclusions to which such experiments lead us are embodied in three statements, generally known as the laws of motion, and first given

by Newton. As enunciated by him these laws are as follows:

LAW I. *Every body will continue in its state of rest or of uniform motion in a straight line, except in so far as it is compelled by impressed force to change that state.*

LAW II. *Change of motion is proportional to the impressed force, and takes place in the direction in which that force acts.*

LAW III. *Action and reaction are always equal and opposite.*

40. The evidence upon which these laws are accepted may be stated as follows:

In the first place, our daily observations of phænomena around us, and the results of rough experiments, lead us to infer the probable truth of the principles enunciated in the statement of the laws; it is then found that the more nearly we make the conditions, which obtain in our experiments, approximate to the ideal conditions of the case to which the laws are immediately applicable, the more nearly do the results of our experiments coincide with the principles laid down in these laws; and lastly, the results of long and complicated calculations based on the assumption of their truth are exactly in accordance with natural phænomena. This last evidence is that on which we chiefly rely, and it amounts almost to an absolute proof of the points at issue. Our acceptance of all other physical laws rests on precisely the same kind of evidence. As an example we may refer to the moon, whose motion is calculated on the assumption of the truth of the laws stated above, and of the law of gravitation; and, notwithstanding the extreme complexity of the calculation, we are enabled to determine the moon's position at any instant for years in advance with such precision as to be within the limits of error of the most exact instruments. It was also by the assumption of the truth of these same laws that Prof. Adams and M. Le Verrier were enabled to calculate the position and orbit of the planet Neptune before it had been seen.

41. We shall now consider each of these laws separately, and trace them into some of their consequences.

LAW I. *Every body will continue in its state of rest or of uniform motion in a straight line, except in so far as it is compelled by impressed force to change that state.*

This law supplies us in the first instance with the definition of *force* given above. For if a body do not continue in its state of rest or of uniform motion in a straight line it must be under the action of force; so that force is that which tends to change a body's state of rest or of uniform motion in a straight line. Secondly the law indicates a mode of measuring time. The velocity of a body is uniform when the body passes over equal distances in equal intervals of time. Suppose there are two bodies A and B and that no force acts upon either of them which has a tendency to change its motion relative to the other. Then by the first law of motion B must move uniformly in a straight line relative to A. Hence the intervals of time during which B moves over equal distances relative to A must be equal. Therefore when applied to this system of only two bodies, the first law of motion simply defines equal intervals of time as those during which B moves over equal distances relative to A, but it states no law of nature. Now suppose a third body C introduced and consider its motion relative to A. If C move in a straight line relative to A and pass over equal distances in the intervals during which B passes over equal distances, then B being our time-keeper, it follows that C moves uniformly relative to A and is therefore, by Law I., under the action of no force which has a tendency to change its motion relative to A. If, however, C do not fulfil this condition, some force must act upon it in such a way as to change its motion relative to A. The statement that if both B and C be under the action of no forces tending to change their motion relative to A, then C will move over equal distances in the same intervals during which B moves over equal distances, is one which considered *à priori* might or might not have been true, and it is in this statement that the *law* consists.

FIRST LAW OF MOTION. 27

42. If we were furnished only with the system consisting of A, B, and C, we might have no better reason for supposing B to be under the action of no force, than for supposing C to be free from the action of force. In this case the choice of B as our time-keeper would be a purely arbitrary choice. But suppose there were a large number of bodies B, C, D...all moving relative to A, and suppose that of this multitude a certain number L, M, etc., all moved over equal distances relative to A in the same intervals during which K moved over equal distances. Then all these bodies K, L, M, etc., agree in indicating the same intervals of time as equal intervals, that is, they all provide us with the same measure of time, and if there be no apparent reason why all these bodies should be similarly acted on by forces, we have very good reason to believe that each of them is under the action of no forces and to accept their joint testimony as the basis of our measurement of time; while if the rest of the bodies B, C, D, etc., neither agree with K, L, M, etc., nor yet among themselves, we have good reason to believe that these are acted upon by forces, and to reject them all as means of measuring time. It is from the contemplation of such a system of bodies that our highest conception of the measurement of time is derived, and though we sometimes speak of equal intervals of time as those during which the earth turns through equal angles, we really ultimately refer to the joint testimony of all the heavenly bodies, after duly allowing for all the forces which we know to act upon them, in determining our measurement of time, and thus we may ask whether the length of the day (or the rate of rotation of the earth) is the same now as formerly, a question which would be absurd if our fundamental notions of the measurement of time were based upon the earth's rotation simply. (See Maxwell's *Matter and Motion*.)

43. The first law of motion attributes to matter the property known as inertia, by virtue of which a finite force acting during a finite time is required, to produce a finite change in the velocity of a finite quantity of matter. In other words, it states that any particle of matter

has no power in itself of changing, of its own accord, any velocity with which it may be moving. Now that a body, if at rest, will continue so, if no force act upon it to disturb it, every one will at once admit as in accordance with every-day observation; but that a body in motion will continue to move uniformly in a straight line is not quite so obvious, because we never have an opportunity of observing the motion of a body under the action of *no* forces. If a stone be projected along a horizontal plane it will at length come to rest, but if the stone and plane be made smoother the stone will continue longer in motion, and this leads us to believe that if all opposing forces were removed, the stone would never come to rest. If, however, we reflect on the facts that we have no notion whatever of absolute motion; that all motion which comes under our notice is essentially relative; that a particle appears at rest when it is moving in the same direction and with the same velocity as ourselves; and that we have no means of ascertaining whether a particle is absolutely at rest or not, we see that there is no distinction in kind between uniform motion in a straight line and absolute rest, the latter being in fact only a particular case of the former. The assertion, therefore, that a particle under the action of no forces will remain at rest is entirely without meaning, since we never know whether a particle is absolutely at rest or not, unless it be also true that a particle under the action of no forces will continue to move uniformly in a straight line if it be once in a state of motion, so that this latter part of the law is a necessary consequence of the former.

44. Law II. *Change of motion is proportional to the impressed force, and takes place in the direction in which that force acts.*

The phrase "change of motion" here is equivalent to change of quantity of motion or of momentum. If the force be finite it will require a finite time to produce a sensible change of motion, and the change of momentum produced by it will depend upon the time during which it acts. The change of motion contemplated must then be understood to be the change of momentum produced per unit of time, or the *rate* of change of momentum.* If the force be variable, the rate of change of momentum is measured at any instant by the momentum which would be generated in the unit of time if the force remained constant during that unit and the same as at the proposed instant, and this the law asserts to be proportional to the intensity of the force at that instant. Now the momentum of a moving particle is the product of the mass into the velocity; if, then, the mass remain constant the change of momentum is measured by the product of the mass into the change of velocity, and the rate of change of momentum by the product of the mass and the rate of change of velocity. But rate of change of velocity is acceleration; therefore this law states that any force P acting on a particle of mass m is proportional to the product of the mass m on which it acts, and the acceleration f produced therein by the force;

$$\therefore P \infty\, mf.$$

Hence we may put P equal to kmf, where k is some constant.

45. If the unit of force be taken as that force which produces the unit of acceleration in the unit of mass, or, which acting on the unit of mass for the unit of time generates therein the unit of velocity, then, if we put m equal to unity, that is, take the unit of mass, and f equal to unity, that is, introduce the condition that the acceleration produced therein shall be the unit of acceleration, we must have the force producing the acceleration equal to the unit of force, or P equal to unity. Hence k must also be equal to unity, and we have the equation

$$P = mf.$$

The meaning of this equation is as follows: *The number of units of force in any force is equal to the product of the number of units of mass in any particle on which it may act, and the number of units of acceleration produced in that mass by the force in question.*

* Newton made the second law of motion refer to the *whole change of momentum* produced, which is the equivalent of the *im-*

It should be carefully noted that the unit of force implied in the above equation is *that force which acting on unit mass for unit time produces unit velocity.* This unit of force is implied in all dynamical equations in which no numerical constant is introduced as a factor to change the unit.

46. Suppose that a second and a foot are the units of time and length respectively, a pound being the unit of mass, and that we require to know the corresponding unit of force, in order that the above equation may be true. We may argue thus:

The unit of force in this case is that force which acting on the mass of a pound for one second generates in it a velocity of one foot per second; now we know from the results of experiments, some of which will be described hereafter, that a force equal to the weight of a pound in London, at the sea level, if acting on the mass of a pound generates in it in one second, if free to move, a velocity of nearly 32·2 feet per second; and hence the unit of force is $\frac{1}{32\cdot 2}$ of the weight of a pound, or rather less than the weight of half an ounce. This is the British absolute dynamical unit of force, and is called a *poundal*. In order that the equation $P = mf$ may be universally true when a pound, a second, and a foot are the units of mass, time, and length respectively, all forces must be expressed in terms of this unit.

47. If a centimetre be taken as the unit of length, and a gramme as the unit of mass, a second being the unit of time, the unit of force is that which acting on a gramme for a second produces a velocity of a centimetre per second, and is called a *dyne*. The weight of a gramme at the sea level in Paris is equal to about 981 dynes.

48. Again, from the equation $P = mf$ we see that a force is *measured* dynamically by the *momentum* which it generates in the unit of time. We have thus a method of comparing the magnitude of one force with that of another, and are enabled to express any force in terms of the

pulse of the force; but the interpretation given in the text is more convenient.

IMPULSE OF A FORCE. 31

absolute unit above defined, or of any other unit we may adopt. In fact, the second law of motion provides us with an absolute measure of force.

Suppose P units of force to act uniformly for t seconds on m units of mass, producing f units of acceleration. Then $f = \dfrac{P}{m}$. Also if v units of velocity be produced in the end we have $v = ft = \dfrac{P}{m}t$, or $Pt = mv$. Hence the product of the force into the time during which it acts is numerically equal to the momentum produced by it.

Hence the second law of motion enables us to assign a physical meaning to momentum. It is the effect or product of a force acting for an interval of time. A given force acting for a given time will always generate the same amount of momentum, whatever be the mass upon which it acts, and this momentum is measured by the algebraical product of the measure of the force and of the time during which it acts.

49. If in the interval during which a force acts its magnitude vary, the acceleration produced by it will vary proportionally. If we split up the time of action of the force into intervals so short that we may consider the force as uniform during each interval, the momentum generated during any interval τ during which the force is equal to P units will be equal to $P\tau$, and so for every interval. Therefore the whole momentum generated may be found by multiplying the number of seconds in each interval by the measure of the force, supposed constant during that interval, and adding the results. If the number of intervals be increased and their length diminished indefinitely, we arrive at the case of a continuously varying force. Those acquainted with the Integral Calculus will see that the above may be shortly expressed by saying that the whole momentum generated by a force is equivalent to the "time-integral" of the force itself.

DEF. *The whole momentum generated by a force is called the impulse of the force.*

50. If we know the impulse of a force and the time

during which it acts, we have only to divide the impulse by the time in order to obtain the measure of the force, supposing it uniform, or the time-average of the force, supposing it variable. Sometimes a finite momentum is generated in a time so short we are unable to measure it. In this case the force producing it must be very great, but we cannot tell how great because we do not know the time during which it acts. We know the whole momentum generated, that is, the impulse of the force, but not the force itself. In other cases we may be able to find the momentum produced and the whole time during which the force acts, but may be unable to trace the variation of the force. In this case we can find the impulse of the force and its (time) average value, but not the magnitude of the force at any particular instant.

51. Until recently it was customary to find the following definitions in works on Dynamics:

A finite force is one which requires a finite time to generate a finite momentum.

An impulse is a force which generates a finite momentum in an indefinitely short time.

An impulse is measured by the whole momentum generated by it.

Now we have no experience of a finite momentum being generated in an indefinitely short time, and an infinite force would be required to produce such an effect. Moreover, if an impulse is measured by the whole momentum generated by it, it is something quite different from force, being of one dimension higher in time. According then to this system there were two classes of forces; *viz.* finite forces and impulses, which could not be compared with one another, being, in fact, of different dimensions.

52. The only mode of measuring force now recognised is that indicated by the second law of motion. If the whole momentum generated by a force, or the impulse of the force, be given, then if the force take a long time to generate this momentum the force itself must be correspondingly small; if it take a very short time the force

IMPULSE OF A FORCE. 33

must be correspondingly great. If we cannot measure the time we cannot determine the force, and have then to content ourselves with knowing the momentum generated by it, or its *impulse*, a term invented simply to hide our ignorance of the force itself.

A force considered only with reference to the whole momentum generated by it during its action is called an *impulsive force*. Generally the term impulsive force is applied only to forces which act for a time so short that no sensible changes take place in the configuration of the system during their action, and hence in considering their effect we need take no account of forces which require a much longer interval to produce any appreciable effect.

This is, in fact, the only practical point in distinguishing between impulsive forces and other forces. In the case of impulsive forces it is generally understood that their time of action is so short that no change takes place in the position of the body acted upon, and no effects are produced by the other forces acting on the body while the impulsive forces last, so that, in considering the effects of these forces, we do not complicate the problem by the introduction of other forces which are much less intense but more continuous. Examples illustrating this distinction will be met with in some of the problems relating to the motion of bodies connected by a string and in the chapter on Collision.

53. Suppose a force to generate a finite momentum in $\frac{1}{1000}$th of a second. The force may at first increase, then reach a maximum and subsequently diminish, vanishing at the end of the above-mentioned interval. Except by very refined methods we should be quite unable to measure, even approximately, the time during which the force acted, and could therefore form no idea even of the average magnitude of the force, much less could we detect its variation, and we should have to speak simply of its impulse. But suppose our faculties or instruments so much improved that we could appreciate $\frac{1}{1000000}$th of a second; then we might not only measure the time during which the force acted, but determine its mean

G. D.　　　　　　　　　　　　　　　　　　　　　D

value for each separate millionth of a second during its action. The force would, in fact, be as completely subject to our measurement as a force which acts for a quarter of an hour would be to an observer who could measure and appreciate no interval of time less than a second. Measurements of this character have been actually carried out by Sir Frederick Abel and Captain Noble in their determination of the velocity of a shot at different positions in a rifled gun, and their deduction of the pressure which must be exerted upon it in the several positions by the products of combustion of the powder. An important practical result of these experiments was the determination of the form of the rifled grooves that the twisting couple upon the shot might be uniform, and thus the wear of the groove uniform notwithstanding the variation of the pressure of the powder gases. If the grooves were regular helices, it is clear that the velocity of rotation of the shot would be proportional to the linear velocity, and hence the twisting couple would be always proportional to the pressure of the powder gas and the grooves would wear most in the part of the barrel where the pressure was greatest. By making the pitch of the screw at all points proportional to the pressure of the powder gas the twisting couple is kept constant, notwithstanding the great variation of pressure in the barrel. Of course other very important objects were the determination of the requisite strength of the gun at different portions of its length, and the adaptation of the powder to the character of the projectile.

54. Suppose several forces to act at once on a particle either at rest or in motion; then, the second law of motion being true for *every one* of these forces, and the effect of a force being completely determined by that law, it follows that each must have the same effect, in so far as the change of motion produced by it is concerned, as if *it* were the only force in action; we may therefore infer that

When any number of forces act simultaneously on a body, whether at rest or in motion, each produces the same change in the body's motion as if it alone acted on the body at rest.

This statement expresses the principle of the "*physical independence of forces.*" From it, taken in conjunction with Newton's second law of motion, it follows that if any number of forces act on a particle, initially either at rest or in motion, the equation $P = mf$ will be true for each, f being the acceleration produced by the force P, and this acceleration will be in the direction in which P acts.

From this result, coupled with the "parallelogram of accelerations" proved above, the "parallelogram of forces" immediately follows, and we have at once the whole subject of Statics.

Precisely the same remarks apply to impulses, which may therefore be resolved and compounded in the same way as forces.

55. In all the preceding cases, change of velocity must be estimated in accordance with the parallelogram law. Thus, if a straight line AB represent in magnitude and direction the initial velocity of the particle, and AC its final velocity, BC will represent the whole change of velocity produced.

56. When it is stated in the second law of motion that "change of motion is proportional to the impressed force," it is not implied that the whole change of motion should be that of a *single* particle, or even of a single body, in the usual sense of the term; nor is it necessary that all the changes of motion in the different parts of a complicated system, which act on each other by means of their connections, should be at any instant in the same direction. The change of motion of each *particle* of the system will be in the direction of the resultant of all the forces acting upon it; but some of these forces are introduced by the connections of the system, and the direction of the resultant force on the particle depends on the nature of these connections. Since, however, action and reaction between the particles of the system must be equal and opposite it follows that the resultant momentum generated in the system, as determined by compounding the momenta of all the particles according

to the parallelogram law, is proportional to the resultant of all the forces acting on the system and is the same as this resultant force would produce in the same time on a single particle. This is practically D'Alembert's Principle.

When a system is such that the velocity of one particle being given in magnitude and direction that of all the others can be found, the fact that the resultant momentum of the system corresponds to the effect of the resultant of the forces acting upon it will enable us to determine completely the motion of the system when known forces have acted upon it for a given time. We shall meet with examples of this hereafter.

57. As an application of the second law of motion we may take the following. It has been established as an experimental fact that if two or more bodies of different materials, as, for example, a sovereign and a feather, be allowed to fall in vacuo simultaneously from rest, if let fall from any, the same, height, they will reach the ground together. Hence their velocities at every instant during their fall must be the same. Now since the velocities of all the bodies at *any* subsequent instant are the same, it follows that the accelerations under which they are moving are the same for all. The force acting on each is, therefore, by the second law of motion, proportional to its mass. But this force is that with which it tends to fall towards the earth; in other words, its weight. The weight of each body is therefore proportional to its mass, and *independent of the kind of material of which it is formed*. It follows then, from this experiment, that the earth attracts *all kinds of matter* alike.

58. The acceleration of a body falling freely in vacuo is found to vary slightly with the latitude, and also with the elevation above the sea-level. This acceleration is generally denoted by g, and when we say that at any place g is equal to 32, we mean that the velocity generated per second in a body falling freely under the action of gravity at that place is a velocity of 32 feet per second The value of g at the sea-level in the latitude of Edinburgh

is 32·2 very nearly. It may be mentioned here that a body is said to be moving *freely* when it is acted upon by no forces except those under consideration.

If in the equation $P = mf$ we put for P the weight W of the body, we know that the acceleration produced is g; hence for f we must write g, and we get the equation

$$W = mg,$$

the unit of weight, or of force, being in this case the absolute dynamical unit of force.

59. The *weight* of a body is the *force* with which it tends to move towards the earth, and is equal and opposite to the force which must be exerted in order to support it. It is equal to the attraction of the earth for the body diminished (according to the parallelogram law) by the force which is necessary to cause the body to participate in the diurnal rotation of the earth, and which produces its whole effect in causing acceleration towards the earth's polar axis when the body is at rest relative to the earth's surface.

60. We have seen that the weight of a body is proportional to its mass; therefore bodies whose weights at any given place are equal must contain equal quantities of matter, and we thus obtain a means of determining the equality of two masses by the use of the common balance. Hence, also, we have a practical method of comparing the quantities of matter in two or more different bodies by determining how many masses, each equal to the standard, are required to balance each of the bodies whose masses are to be compared. Sub-multiples of the standard may be obtained at first by weighing out a quantity of matter equal to the standard unit of mass and dividing it by trial into the required number of *equal* parts, the equality of the parts being tested by the balance. In this way a system of " weights " for the measurement of mass may theoretically be obtained.

It should be borne in mind that the ultimate object of weighing things is not generally to ascertain their *weight*, that is, the force with which the earth attracts them, but to determine the quantity of matter contained in them, and this is rendered possible only by the fact that the earth attracts all kinds of matter alike, a fact which we have learned by observing that all bodies fall with the same acceleration in vacuo. The same conclusion also follows from the result of certain experiments on pendulums, which show that, other things being the same, the time of oscillation of a pendulum is independent of the material of which it is composed.

61. The characteristic difference between the results obtained by a spring balance and by a pair of scales is this. A pair of scales is used to determine at once how many units of mass are contained in the body weighed. This is done by determining how many units of mass are attracted by the earth with the same force as the body in question, and since, as above stated, all kinds of matter are attracted alike by the earth, the result is the number of units of mass contained in the body; and this test is entirely independent of the absolute intensity of gravitation, that is, of the value of g, at the place. The spring balance, on the other hand, simply measures the force with which the earth attracts the body weighed; the apparent weight of a body will therefore depend on the intensity of gravity at the place. Now the acceleration produced by gravity in a body falling freely is less at the equator than at the poles, and increases continuously with the latitude; it also diminishes as the altitude above the sea-level is increased. The apparent weight of a body will consequently be less in low than in high latitudes, and a spring balance will be disadvantageous to a merchant buying goods in England and selling them at Cape Coast Castle; while if a pair of scales or a steelyard be used in its place, the apparent weight of the merchandise will be independent of the latitude.

62. DEF. A force is said to do *work* when it moves its point of application.

An agent is said to do work when it overcomes resistance.

Since action and reaction are exactly equal and opposite it matters not whether we consider the work done by an agent as proportional to the force exerted, and the distance through which the point of application is moved in the direction of the force, or as proportional to the resistance overcome and the distance through which it is overcome.

The work done by a force whose magnitude and direction remain constant, is proportional to the product of the intensity of the force into the distance through which its point of application has been moved in the direction of the force.

From this it will be seen that, if the point of application move always in a direction perpendicular to that of the force, the latter does no work. Thus no work is done by gravity in the case of a particle moving on a horizontal plane, and when a particle moves on any smooth surface no work is done by the force which the surface exerts upon it. If, on the other hand, a heavy body be lifted from the ground, the agent raising it does work upon it, and the work done is proportional to the product of the weight of the body and the vertical height through which it is raised. In this case the body is moved in the direction opposite to that in which its weight acts, and the work done by the earth's attraction is accordingly negative. When the work done by a force is negative, that is, when its point of application moves in the direction opposite to that in which the force acts, this is frequently expressed by saying that work is done *against* the force. In the above case work is done *by* the agent lifting the heavy body and *against* the earth's attraction.

63. If the amount of work which is done by the unit of force, when its point of application moves through the unit of length in the direction of the force, be taken as the unit of work, then the measure of the work done by a force whose magnitude and direction remain constant will be the product of the numbers representing respectively the force and the distance traversed by its point of application in the direction of the force. Choosing it so that this condition may be satisfied, the British absolute unit of work is that done by the absolute dynamical unit of force, or poundal, when its point of application moves through one foot in the direction of the force, and is called

a *foot-poundal*. It is equal to the work done in lifting rather less than half an ounce one foot high.

The unit of work generally adopted by engineers is the foot-pound, that is, the work done against gravity by an agent in raising the mass of a pound through the vertical height of one foot. Now the mass of a pound being invariable, its weight varies with the locality on account of the variation of g. The foot-pound is therefore not an invariable standard but depends on the locality, and is consequently unsuited for a scientific unit of work. The foot-pound contains g absolute units of work because the weight of a pound is equivalent to g absolute units of force.

The unit of work belonging to the centimetre-gramme-second system, or the C.G.S. unit of work, is that done by a *dyne* in working through a centimetre and is called an *erg*. It is rather more than the work done in lifting a milligramme against gravity through one centimetre, for since g is equal to about 981 centimetre-second units of acceleration it follows that the weight of a gramme is about 981 dynes and that the work done in lifting a gramme through one centimetre is 981 ergs.

64. When either the magnitude or direction of a force varies, or if both of them vary, the work done by the force during any finite displacement cannot be defined as above. In this case the work done during any indefinitely small displacement may be found by supposing the magnitude and direction of the force constant during that displacement, and estimating the work done in accordance with the above definition : taking the sum of all such elements of work done during the consecutive small displacements, which together make up the finite displacement, we obtain the whole work done by the force during such finite displacement.

The effect or product of a force, when its point of application moves over any distance in the direction of the force, is a certain amount of work, and this work is measured by the algebraical product of the measure of the force and of the distance moved over by its point of application in its direction.

HORSE-POWER.

65. DEF. The *rate* at which an agent works is measured when uniform by the amount of work done by it in the unit of time: when variable it is measured at any instant by the amount which would be done by it in the unit of time, if the rate remained uniform during that unit and the same as at the proposed instant.

The rate at which work is done by a force is the product of the force and the velocity of its point of application in the direction of the force.

DEF. The *power* of an agent is proportional to the rate at which it can work. An agent capable of performing 33,000 foot-pounds of work per minute is said to be of one *Horse Power*. Thus, when we say that the *actual* horse-power of an engine is ten, we mean that the engine is able to perform 330,000 foot-pounds of work per minute. The *nominal* horse-power of a steam-engine depends only on the number and measurements of its cylinders, and the nature of the engine, and not upon the actual rate at which it can work, which varies with the pressure of steam in the boiler and is limited by the strength of the latter. The letters H. P. are often used as abbreviations of the words *horse-power*.

It will be seen that the horse-power, like the foot-pound, is not an absolute unit, but depends on the intensity of the earth's attraction at the place. All such units are sometimes classed together under the name of *gravitation units*.

The British absolute unit of power is that of an agent which can perform one foot-poundal per second.

The C.G.S. unit of power is that of an agent capable of doing one *erg* per second. As this is an inconveniently small unit another unit based on the same system is adopted by electricians for the *rate of doing work*, and is called a *Watt*. The *watt* is 10,000,000 *ergs per second*. It should be noticed that the watt, like the horse-power, is not a *quantity of work*, but a *rate of doing work*, and is of the nature of *work divided by time*. The horse-power is equivalent to 745·8 watts.

66. DEF. The amount of work which a system is cap-

able of doing in passing from its present condition to some standard condition is called its *energy*.

A system may possess energy in virtue of its configuration, or the relative positions of its parts. Thus a distorted spring can do work in returning to its natural form; the system consisting of the earth and a raised weight can do work by their mutual approach. The energy which a system possesses in virtue of its configuration is called *potential energy*.

67. If we catch a cricket ball when the ball strikes the hand it exerts a pressure upon it, and unless the hand be made to move in the direction of the ball's motion it may inflict a serious injury. If the hand be withdrawn somewhat as the ball strikes it, the ball exerts pressure upon the hand while it moves in the direction of the pressure. It therefore does work, and we see that work may be done when two bodies collide, in virtue of their relative motion. In order that work may be done we must have at least two bodies moving relatively to one another, for if the cricket ball be not acted upon by any other body it will go on moving indefinitely and have no opportunity of doing work. Similarly, in order that a system may have potential energy we must have at least two bodies or portions of a body capable of changing their relative positions.

The energy which a system possesses in virtue of the relative motions of its parts is called *kinetic energy*.

68. If a moving body such as a cannon shot strike the earth and come to rest relative to it, it will not sensibly affect the earth's motion, in consequence of the enormous mass of the earth as compared with the shot. In this case we may assume that the shot loses the whole of the velocity which it had relative to the earth, and it will be shown hereafter (Chapter II.) that in such a case the number of units of work done is the same as the number of units of kinetic energy possessed by the shot as defined in Art. (38). Whenever we speak of the kinetic energy of a single moving particle, we mean the kinetic energy possessed by the particle and the earth in virtue of their

relative motion; similarly when we speak of the potential energy of a raised weight, we mean the potential energy possessed by the weight and the earth in virtue of their relative positions.

69. LAW III. *Action and reaction are always equal and opposite.*

If a body A press against another body B with which it is in contact, B will exert an equal pressure on A, but in the opposite direction. But the application of this law is not confined to the mutual actions between surfaces in contact, the statement being true for all kinds of mechanical action whatsoever. Thus, while in accordance with the law of gravitation the sun attracts the earth with a certain force, the earth also attracts the sun with an equal force, and these two attractions are, of course, in opposite directions. Again, the earth attracts a falling rain-drop with a certain force, while the rain-drop attracts the earth with an equal force. The result is that while the rain-drop moves towards the earth on account of its attraction, the earth also moves towards the rain-drop under the influence of the attraction of the latter, but the mass of the earth being enormously greater than that of the rain-drop while the forces on the two arising from their mutual attractions are equal, the motion produced thereby in the earth is all but incomparably less than that produced in the rain-drop, and is consequently quite insensible. The third law of motion is also applicable to electrical attractions and repulsions, to the actions of magnets and of conductors conveying electric currents on themselves and each other, and, in fact, to all cases of mechanical action.

70. The first law of motion states the property of *force* and gives us, so to speak, a *qualitative* test of its presence, for from it we infer that when a body changes its state of rest or of uniform motion in a straight line it is under the action of force.

The second law of motion raises force to the dignity of a mathematical *quantity* and explains how it is to be measured. Combining this law with a purely *kinematical* theorem, viz. the parallelogram of accelerations, we learn how to find the resultant of any number of forces, and the subject of *statics* follows.

The third law of motion is a brief summary of a number of phenomena having a general resemblance but differing in their details, and is intended rather to be a conventional mode of expressing in as few words as possible a series of experimental results, than to *connote* the several physical phenomena which it includes.

71. If one body attract or repel another, the second attracts or repels the first with an equal and opposite force. This is true not only of the attractions of the heavenly bodies but of molecular forces between the particles of matter constituting any solid, liquid or gas. (Were this not the case we might have a body A attracting another body B with a force greater than that with which B attracts A. Then considering A and B together as forming one system there would be a *resultant* force upon the system acting from B towards A and making its centre of inertia move with an acceleration from B to A, while no force is applied by external agency, and this is contrary to experience.)

If a body A press against another body B both bodies being at rest, then every one will at once admit that B presses A with a force equal and opposite to that which A exerts upon B. Thus if a pound weight rest on a horizontal table we know that the table exerts a vertical pressure on the pound equal to the weight of a pound, because it supports it, and no one will doubt that the pressure of the weight on the table is equal to the weight of a pound, (in fact, some persons will consider the latter statement more obvious than the former,) so that in this case action and reaction are equal and opposite.

72. Suppose the finger pressed against a piece of soft putty or other material so as to penetrate it; the question may be asked—"Is the pressure of the putty on the finger in this case equal to the pressure of the finger on the putty? and if so why does the finger penetrate the putty?" Consider a very thin section of the finger

THIRD LAW OF MOTION. 45

which includes all the portion in contact with the putty; let m denote the mass of the section and f its acceleration towards the interior of the putty. Then the *resultant* force on the section is mf towards the putty, and the pressure exerted on the section by the rest of the finger must be greater than the pressure of the putty upon it by the quantity mf. But by taking the section of the finger sufficiently thin we can make m as small as we please. Hence the pressure exerted by the putty upon an *indefinitely* thin section of the finger in contact with it is indefinitely nearly equal to the pressure exerted by the rest of the finger upon that section, and this latter pressure differs by the indefinitely small quantity, mf, from the pressure exerted by the finger on the putty. In this case, then, the action and reaction between the finger and the putty are equal and opposite.

If the pressure of the hand on the knuckle end of the finger be greater than that which the putty can exert upon the other end, there will be a resultant force upon the finger towards the putty, and this force will produce acceleration in the finger which will therefore penetrate the putty with an accelerated motion, and the pressure of the finger upon the putty will be equal only to the pressure exerted by the putty on the finger, and not equal to that exerted by the hand on the knuckle end of the finger, the difference being that force required to produce acceleration in the finger, so that action and reaction between the *end* of the finger and the putty are equal and opposite. If the pressure exerted by the hand on the finger be *equal* to that exerted by the putty upon it the finger will remain at rest or penetrate the putty with *uniform* velocity.

73. Suppose two particles whose masses are respectively m and m' to be connected, and the particle of mass m to be acted on by a force P in the direction of the line joining the two. The acceleration of two particles will be the same as that of one particle of mass $m + m'$, and will therefore be denoted by $\dfrac{P}{m+m'}$. Now the force required to produce this acceleration in the particle of mass m' is

$\dfrac{Pm'}{m+m'}$, and this is therefore the force with which the first particle acts on the second and is in the same direction as P. Also the force required to produce the acceleration $\dfrac{P}{m+m'}$ in the first particle is $\dfrac{Pm}{m+m'}$, and this therefore represents the *resultant* force upon it. But one of the forces applied to the particle is P, and the other is the reaction of the second particle; this latter is therefore in the direction opposite to that of P and equal to $P - \dfrac{Pm}{m+m'}$, that is, to $\dfrac{Pm'}{m+m'}$, and is therefore equal and opposite to the action of the first particle upon the second. Hence in this case action and reaction between the particles are equal and opposite.

We see then that in all cases in which force is exerted between *portions of matter*, whether at rest or in motion, action and reaction are equal and opposite, and this is as true for the molecular forces which act between the ultimate molecules of matter as for the forces of gravitation between the heavenly bodies.

74. If m denote the mass of a particle, and f its acceleration, the force mf required to produce this acceleration is called the *effective* force on the particle, and is identical with the resultant of all the forces acting on it.

Imagine any connected system, and let m denote the mass of one of its particles and f its acceleration. Then the effective force on the particle will be denoted by mf, and this must be equivalent to the resultant of all the forces acting on the particle, and will therefore if reversed maintain equilibrium with them; and this is true for *all* the particles in the system. Therefore the reversed effective forces of *all* the particles will balance all the other forces acting throughout the system. But the forces applied to any particle consist in the most general case of two classes; viz. those impressed upon it by external agency, and the forces exerted by the other parts of the system which we may call *internal* forces. But to each force between the parts of the system there is (by the third law of motion) an equal and opposite reaction, so that the internal forces taken *throughout the system* are in equilibrium amongst themselves. Hence the effective forces of all the particles when reversed will maintain equilibrium with the forces impressed from without. This is D'Alembert's principle and it follows at once from Newton's laws of motion.

All that has been said above is true however great the forces may be, or however small the time during which they act, and will therefore apply to forces which act only during an interval too short for measurement.

75. It must be remembered that the action and reaction contemplated in the third law of motion are action and reaction between *portions of matter*. There is nothing in nature corresponding to a "force of resistance against acceleration" which is supposed by many writers to be exerted by matter, and it cannot be too strongly urged that reversed effective forces have no real existence. But Newton's third law of motion admits of a wider interpretation than this. In the scholium to this law Newton says:—

If the action of an agent be measured by its force and velocity conjointly; and if similarly the reaction of the resistance be measured by the velocities and amounts of its several constituents conjointly, whether these arise from friction, cohesion, weight, or acceleration;—action and reaction in all combinations of machines will be equal and opposite.

Now it has been shown that the product of a force into the velocity of its point of application in the direction of the force is the *rate* at which it *works*. Newton then in this scholium measures the action of an agent by the rate at which it works, and similarly he measures the reaction of the resistances by the *rate at which work is done against them*. The work done against the resistance arising from acceleration mentioned in the scholium is the work done by the *effective* forces of the system on account of their *producing acceleration* and is expended in generating kinetic energy in the system. Hence the measure of the reaction arising from acceleration is the rate at which kinetic energy is being generated in the system, and the scholium when interpreted into modern phraseology will stand thus:—

If the action of an agent be measured by the rate at which it works, and similarly the reaction of the resistances arising from friction, cohesion, and weight by the rate at which work is done against them, and if we include amongst the measures of these reactions the rate at which kinetic energy is being generated in the system;—action and reaction in all mechanical combinations will be equal and opposite.

76. The statement thus interpreted is nothing more nor less than the enunciation of the great principle of the conservation of energy. If an agent work upon a system and there be no opposing forces such as friction, etc., then the rate of change of the kinetic energy of the system is precisely equivalent to the rate at which the agent works, and therefore the whole change of kinetic energy produced in the system in any time is equivalent to the work done on the system by the agent.

If, however, the action of the agent be opposed by forces of the nature of friction, etc., the rate of change of kinetic energy in the system is less than the rate at which the agent works by the rate at which work is being done against these opposing forces, and the whole change of the kinetic energy of the system produced in any time is less than the work done by the agent in that time by the work done against friction, etc., and converted into heat or other forms of energy. Together with friction must be classed all forces whose action does not remain constant in magnitude and direction whether the system be at rest or moving in any way whatever.

77. When the agent works against forces of the same nature as weight, with which we include all those which are independent of the time, and of the velocity of the system, depending only on its position and configuration, and which are sometimes called *conservative* forces, then the rate of change of kinetic energy in the system is less than the rate at which the agent works by the rate at which a work is done against these other forces, and

the work done against these forces in any time becomes *potential energy* in the system, and can be converted into kinetic energy by leaving the system free to return to its original position and configuration.

If friction, or other forces of like nature, act on a system, then, when the motion of the system is reversed, these forces are also reversed, and the work done against them is not converted into kinetic energy on leaving the system free to return to its original configuration, having been at first converted into heat, sound, the energy of electric *currents*, or other forms of energy.

78. We see then that Newton's third law of motion consists of two distinctly different principles. The first states the equality of the action and reaction between portions of matter (and is equally true of electrified and magnetized matter), in which case both action and reaction are of the nature of forces such as those considered in statics, or as a particular case of these, of great forces acting for a short time, and the effects of which we find convenient to treat as impulses. The second consists of the great principle of conservation of energy, and asserts the equality of action and reaction when these terms do not imply statical forces but rates of doing work; that being called action *by* which work is done and that *against* which work is done being treated as a reaction, the rate of increase of kinetic energy being included amongst the latter in the statement that action and reaction are equal and opposite.

The endeavour to make the law apply in its first sense to the action of a force on a particle, free to move, was the origin of the introduction of a force of resistance to acceleration equivalent to the effective force reversed, to which there is nothing in nature at all corresponding.

79. Many modern writers, following out the suggestions of the late Dr. Whewell, have enunciated the principle of " the physical independence of forces " as their second law of motion, and Newton's second law they have called the third law of motion. The third law, as given by Newton, they have then either assumed as an axiom or treated as a fourth law. The prevailing tendency amongst

G. D.   E

THIRD LAW OF MOTION.

mathematicians has recently been to return to the laws as enunciated by Newton, and in this form they have been given above.

80. If a particle of mass m be acted upon by a force P which would cause it to move, were it free to do so, with an acceleration f, and be prevented from moving by a string or some other means of constraint, then the string or other constraint must exert on the particle a force equal to P, but in the opposite direction; and, since action and reaction are equal and opposite, it follows that the particle will exert a force upon the string or other constraint equal to, and in the same direction as, the force P. Thus if a heavy particle be suspended in equilibrium by a vertical string, it exerts a downward force on the string equal to its own weight. Similarly a heavy body at rest on a horizontal table exerts a pressure on the table vertically downwards, and equal to the weight of the body.

81. If the particle whose motion is constrained by a string, or other means of constraint, be moving with a velocity uniform in magnitude and direction, since the motion of the particle is not accelerated there can be no *resultant* force acting upon it, and the conditions are precisely the same as if the particle were at rest.

If the particle acted on by the force P, instead of moving with an acceleration f, be constrained by some other means to move with an acceleration f_1 in the direction in which P acts, we can find immediately the force exerted by the constraint. For since the force P would produce in a particle of mass m an acceleration denoted by f, we must have P equal to mf. Also, the particle moves in the direction in which P acts with an acceleration f_1; the resultant force on it must therefore be in the direction of P, and numerically equal to mf_1. The force exerted by the constraint must therefore be in the direction in which P acts, and algebraically represented by $mf_1 - mf$, or by $m(f_1 - f)$. If f_1 be less than f, this shows that the force exerted by the constraint is in the direction opposite to that in which P acts, and numerically represented by $m(f - f_1)$.

If the actual acceleration of the particle be in the direction opposite to that in which P acts, it will be of negative sign. Suppose it to be $-f_1$. Then the force exerted by the constraint must be in the direction opposite to that of P, and numerically equal to $m(f_1+f)$, that is to mf_1+P.

If the actual acceleration of the particle be not in the straight line in which P acts, then we may find the resultant force which must act on the particle in order to produce this acceleration, and the force which is exerted by the constraint is that which must be compounded with P to produce this resultant. This can be found at once by help of the parallelogram of forces.

82. As an example of the preceding article, suppose a heavy particle to rest on a horizontal plane, while the plane moves vertically downwards with an acceleration f; what will be the pressure of the particle on the plane, the acceleration produced by gravity in a particle falling freely being denoted by g?

The pressure of the particle on the plane must, by the third law of motion, be equal and opposite to that of the plane on the particle, and the only forces acting on the particle are the earth's attraction and the pressure of the plane. Now the particle moves downwards with an acceleration f; the *resultant* force upon it must therefore act downwards and be numerically equal to mf. But the attraction of the earth upon the particle acts downwards and is numerically equal to mg. Hence if f be less than g, the pressure of the plane on the particle acts upwards and is equal to $mg-mf$, that is, to $m(g-f)$; while the pressure of the particle on the plane, which is equal and opposite to this, acts vertically downwards and is numerically represented by the same expression, viz. $m(g-f)$.

If f be greater than g, the downward force on the particle must be greater than its weight; the pressure of the plane on the particle must therefore act downwards, and its magnitude will be represented by $m(f-g)$. The particle must therefore be underneath the plane, and its pressure against the plane will be directed upwards.

If f be equal to g the particle is falling freely, and the pressure on the plane becomes zero as we should expect.

83. Suppose, for example, the acceleration produced by gravity in a particle falling freely to be denoted by 32, and a horizontal plane to move downwards with an acceleration denoted by 16: then the pressure on the plane of a heavy particle resting upon it will be one half the weight of the particle. Also, if a heavy body be suspended from a spring balance, and the body and balance together be moved with an acceleration downwards, the apparent weight of the body will be diminished.

By precisely similar reasoning it may be shown that if a horizontal plane on which a particle rests be made to move with an acceleration upwards denoted by f, the pressure of the particle on the plane will exceed its weight by mf units of force; and if a mass m be suspended from a spring balance which is made to move with an upward acceleration denoted by f, its apparent weight will be increased by the quantity mf.

84. The following example will illustrate the subject of the preceding article.

A balloon is ascending vertically so that a pound weight presses on the hand of the aeronaut sustaining it with a force equal to the weight of seventeen ounces. Find the acceleration of the balloon, g being supposed equal to 32 when a second and a foot are the units of time and length.

The forces acting on the pound are a force equal to the weight of 17 oz. acting upwards, and its own weight, that is, the weight of a pound, acting downwards. The resultant force is therefore equal to the weight of an ounce acting upwards. Now the weight of a pound acting on the mass of a pound produces in it an acceleration denoted by 32; therefore, by the second law of motion, a force equal to the weight of an ounce will produce in the mass of a pound an acceleration denoted by 2. Hence the upward acceleration of the pound weight, and therefore that of the balloon, is such as to generate in one second a velocity of two feet per second.

85. Before quitting this portion of our subject we will

take one more example, which will be useful to us hereafter.

Two weights P *and* Q, *whose masses are* M *and* m *respectively, are connected by a weightless string which passes over a smooth pulley. Supposing* M *to be greater than* m, *find the acceleration of each and the tension of the string.*

Let T denote the tension of the string. The mass of P has been denoted by M; its weight is therefore equal to Mg. Also the weight of Q is equal to mg. The resultant force acting on P is consequently $Mg - T$ acting downwards, and the acceleration produced by this is, by the second law of motion,

$$\frac{Mg - T}{M} \text{ or } g - \frac{T}{M}.$$

Again, the resultant force acting upwards on Q is $T - mg$. The upward acceleration produced in Q by this force is

$$\frac{T}{m} - g.$$

Now, since the string remains of invariable length and

always tight, the downward velocity of P is *always* numerically equal to the upward velocity of Q, so long as the conditions of the motion remain unchanged. Hence the changes of their velocities in any interval, and therefore their accelerations, must be numerically equal.

Therefore $\quad g - \dfrac{T}{M} = \dfrac{T}{m} - g;$

$$\therefore T = 2g\,\dfrac{Mm}{M+m};$$

and the acceleration of each weight is represented numerically by

$$g\,\dfrac{M-m}{M+m}.$$

The actual acceleration of P and Q is therefore that which would be produced in a particle whose mass is the sum of the masses, moved by a force equal to the difference of the weights of P and Q; that is, to the force available for producing motion in the system.

86. The results of many experiments may be cited in corroboration of the truth of Newton's second law of motion. We shall describe one piece of apparatus known as Attwood's machine, and show how the results of experiments made with it bear upon this law and other points already mentioned.

Attwood's machine consists of a vertical pillar AB carrying a light brass pulley C; the axle passing through the centre of C turns, with the pulley, and is supported on the circumferences of four light brass wheels, called friction wheels, of which only two, a and b, are shown in the figure. As the pulley C turns, its axle rolls on the circumferences of the four friction wheels, causing them to turn very slowly, their axles working on fixed "centres" supported by the pillar AB. By this means the resistance to the motion of the pulley C, due to friction, is very much diminished, since the only sliding frictions introduced are those against the axles of the friction wheels, and these turn very slowly indeed compared with the pulley C. A light string, to the ends of which two equal weights P and Q are attached, passes over the pulley C. SF is a vertical graduated bar upon which the ring D is

ATTWOOD'S MACHINE.

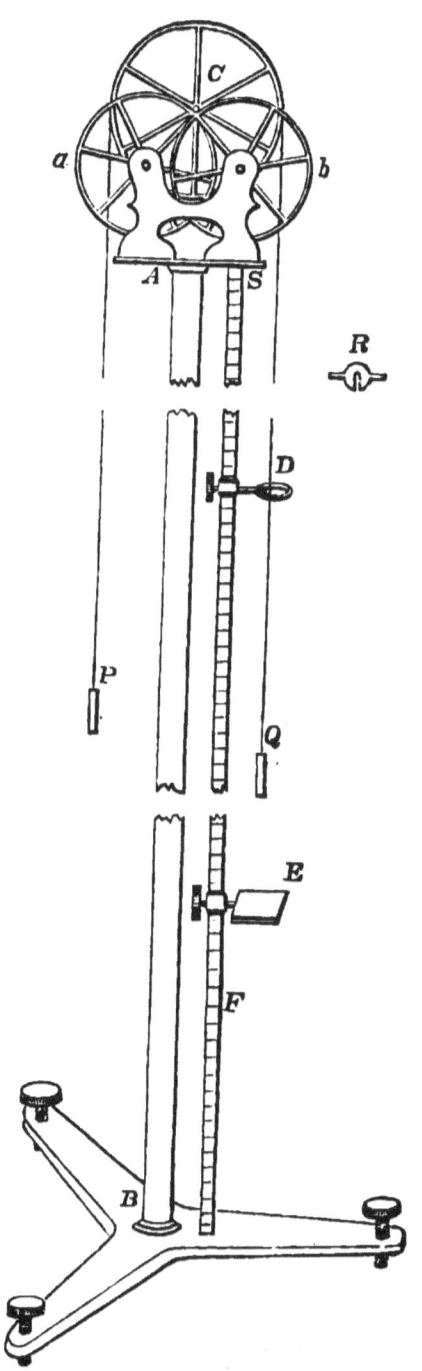

capable of sliding, and of being fixed at any position by means of a set screw; the ring being of such size as to allow of the weight Q passing freely through it. E is a platform also moveable on the bar SF, and, like the ring D, capable of being fixed upon it in any position. R is an elongated weight which may be placed on the top of the weight Q, but which is incapable of passing with it through the ring D. The bar SF may be graduated in feet and hundredths of a foot, or in accordance with any other convenient system. A clock whose pendulum beats seconds is used for timing the experiment, and for the sake of exactitude and convenience other appendages are attached to the instrument, such as an apparatus for suddenly releasing the weight Q, an electro-magnet, which when excited presses a pencil against a cylinder covered with paper and caused to rotate uniformly by the clock, the connection of the coil of the electro-magnet with the battery being made and broken by the motion of Q, or some other form of electric chronograph, etc.; but the parts described above form all the *essential* portions.

Suppose the weight Q to be raised above the ring D, and the weight R placed upon it, and that at a particular instant the system is left free to move. Then since the weight Q is equal to P the weight of Q and R together is greater than that of P; they will therefore descend, and P will be raised. The time required for the top of Q to reach the ring D is then accurately measured, and as the top of Q passes through this ring the weight R is removed. Then, since the weights of P and Q are equal, there is no force to change the motion of the system, which will therefore continue uniform until the weight Q strikes the platform E. The time elapsing between the instants when the weight R is removed from Q and when Q strikes the platform E is accurately observed, and the distance between the ring D and the top of Q when resting on the platform is measured by the graduations on the bar SF. Hence the velocity of Q between D and E can be at once found.

We shall now describe in detail a few experiments with Attwood's machine which have an important bearing on some of the foregoing articles.

EXPERIMENTS WITH ATTWOOD'S MACHINE. 57

87. Exp. I. Let the equal weights P and Q be formed of the same material, M being the mass of each, and let m be the mass of R. Let h be the distance between the ring D and the top of Q when resting on the platform; t the time elapsing between the commencement of the motion and the instant when the weight R is removed from Q, and t' the interval between this instant and that at which Q strikes the platform E.

Then if the same weights be used, and the distance between D and E kept constant, while the depth of the ring D below the point at which Q is liberated is varied so as to vary t, it is found that t' varies inversely as t. Now the velocity of P or of Q after the latter has passed through D is constant, and therefore equal to $\dfrac{h}{t'}$. Hence the velocity varies as t; that is, the velocity generated in the moving system by the weight of R is proportional to the time during which this weight acts upon the system. But the velocity generated by a force is proportional to the time during which that force acts only when the acceleration produced by it, and therefore the force itself, is constant, the mass moved being supposed to remain unaltered. Therefore the force acting upon the system, that is, the weight of a mass m (since m is the mass of R), is constant and independent of the velocity with which the mass is moving. This being found by experiment to be true for any body used in place of R, it follows that the earth's attraction on any mass of matter, that is, its weight, is independent of the velocity with which it is moving. We may therefore introduce a symbol w to represent *throughout the motion* the weight of the mass m, that is, of R.

88. Exp. II. If we neglect the masses of the string and of the pulley C, the whole mass moved is $2M + m$, and the force producing the motion is w. Now suppose the weights P and Q to be changed, P remaining always equal to Q, while R is unaltered. Then $2M$ is changed, and the ratio in which it is changed may be supposed known, for P, Q, and R may be formed by joining together equal weights of the *same* homogeneous matter,

whose masses are consequently equal. By adding to or removing from P and Q the same number of such small equal bodies, the sum of the masses of P, Q, and R, that is, the whole mass moved, may be changed in various *known* ratios. Then it is found from the experiment that the velocity of Q between D and E varies as $\dfrac{t}{2M+m}$. Also, as was found in the preceding experiment, when $2M+m$ remains constant, the velocity of Q after passing D varies as t. Hence the velocity generated in the unit of time by the constant force w varies as $\dfrac{1}{2M+m}$; that is, in the inverse ratio of the whole mass moved.

The result of this experiment agrees with that deduced in Article 85, in which the second law of motion was taken as the basis of the investigation.

89. EXP. III. Suppose, as in the preceding experiment, that P, Q, and R are all made up of a number of equal small masses of the same material, and let any equal numbers of these small weights be removed from P and Q and attached to R. Then w, the force producing motion, is changed in a new ratio; for the weights of each of the small bodies making up R being the same, the weight of R must be proportional to the number of these small weights contained in it. Also, the whole mass moved is the same as before, and it is found that $\dfrac{h}{t'}$, that is, the velocity of the system after Q has passed through D, varies directly as the product wt. Therefore the velocity generated in the unit of time varies as w. But the mass remaining always the same the momentum varies as the velocity; therefore, the *momentum* generated in the unit of time varies as w, the force producing the motion. This result is in accordance with that deduced in Article 85 from the second law of motion, and may be taken as direct evidence in favour of the truth of that law. It also shows that the momentum generated by a force in the unit of time is a proper measure of that force, the unit of force in such case being that force which generates the unit of momentum in the unit of time.

90. Exp. IV. Let the bodies P and Q be exchanged for bodies of other, the same, material, whose *weights*, as determined by a balance, are the same as those of P and Q (*e.g.* brass weights exchanged for iron, platinum, glass, china, or other material), while R remains unchanged, so that the force producing motion remains unaltered. Then experiment shows that the *velocity* generated in the unit of time is independent of the material of which P and Q are composed. But by the second law of motion the *momentum* generated in the unit of time is always the same. Hence, since both the velocity and the momentum generated in the unit of time are independent of the material of which P and Q are made, it follows that the whole mass moved, and, therefore, the masses of P and Q are independent of the kind of that material. But P and Q were taken so that their *weights* should remain unchanged; hence, if the weights of a number of bodies of different material be equal, their masses will also be equal: another proof of the statement that the earth exerts the same attraction on all kinds of matter alike.

91. The first experiment has shown us that the attraction of the earth on a moving body is independent of the velocity with which it is moving. The second, that the momentum generated in the unit of time by a *constant force* is constant and independent of the mass in which it is generated. The third, that the momentum generated in the unit of time in the *same mass* is proportional to the force producing motion. Combining the results of the second and third experiments, we see that the momentum generated in *any* mass in the unit of time is proportional to the force producing motion, and hence we obtain a dynamical measure of force. From the result of the fourth experiment we infer that the earth exerts the same attraction on all kinds of matter, or that the weight of a body depends only on the *quantity* of matter it contains, and is independent of the *kind*, or *quality*, of that matter.

92. In the preceding experiments suppose $\frac{h}{t}$ equal to v. Then v is the velocity of Q when its top passes

through the ring D; that is, the velocity generated in a mass $2M + m$ by the weight of the mass m acting upon it for a time t (see Art. 85); and the experiment shows that this velocity is proportional to $\dfrac{m}{2M+m} t$. We have therefore,

$$v = g \, \frac{m}{2M+m} \, t,$$

g being a constant throughout the series of experiments. Therefore the velocity generated in the unit of time is

$$g \, \frac{m}{2M+m};$$

and since the whole mass moved is $2M + m$, it follows that the momentum generated in the unit of time is mg. But if the unit of force be that force which acting on the unit of mass for the unit of time generates in it the unit of velocity, it has been shown that the measure of any force will be the number of units of momentum generated by it in a given time. Hence the weight of the mass m is represented numerically by mg, or

$$w = mg.$$

The value of g is found to be always the same at the same place, but to vary with the latitude of the place of observation, being greater near the poles than at the equator, and increasing continuously with the latitude. It is also less at great heights above the sea-level than at that level. In the latitude of Edinburgh at the sea-level, when a second and a foot are taken as the units of time and space, the value of g is found to be 32·2 very nearly.

The value of g at the sea-level in latitude λ is approximately represented by the expression

$$G(1 - 0·0025659 \cos 2\lambda),$$

where G obviously represents the acceleration of gravity at the sea-level in latitude $45°$, and is equal to 32·1703 foot-second, or to 980·533 centimetre-second units.

93. We have seen that the momentum generated in a moving system in the unit of time, by the weight of a

mass m, is represented by mg. Now, in the case of a particle falling freely, the force producing motion is its weight while the mass moved is simply its own mass. Hence the velocity generated in the unit of time will be g, which is therefore numerically equal to the acceleration produced by gravity in a body falling freely; and when we say that at a particular place g is equal to 32·2, we imply that at that place the attraction of the earth is such as to generate in a body falling freely a velocity of 32·2 feet per second during each second of its fall. Such motion is called uniformly accelerated motion. This result might of course have been obtained directly from observations on bodies falling freely, but since it is very difficult to measure great velocities, especially when they are continually varying, and since the resistance of the air becomes considerable when the velocity of the moving body is great, Attwood's machine is convenient, inasmuch as the velocity to be actually observed is uniform, and, by sufficiently diminishing the ratio of m to M, can be made as small as we please.

94. In the above explanation of the use of Attwood's machine we have for simplicity neglected the motion of the pulley C. The mass of this pulley is, however, generally comparable with that of P or Q, and the points on the circumference of C have the same velocity as P or Q, while the points within the circumference are moving with less velocity. Were the whole mass of the pulley collected at the circumference we should only have to add this mass to that of the other moving bodies to obtain the whole mass moved; but since that is not the case a complete correction may be made by adding a term M', less than the number of units of mass in the mass of C, to the quantity $2M + m$, and thus employing $2M + M' + m$ as the expression for the whole mass moved. To determine theoretically the value of M' we require some knowledge of Rigid Dynamics; it is sufficient here to state that it admits of exact determination. The motion of the friction rollers is so slow and their momentum consequently so small that they may be altogether neglected.

62 EVIDENCE IN FAVOUR OF THE LAWS OF MOTION.

The error arising from the friction of the pulley can be best compensated by making the two weights, which we have supposed equal, slightly unequal, giving such preponderance to the weight which is to descend during the experiment that when the weight m is removed, the system, if just started, may continue to move with perfect uniformity. If the suspending cord be silk its weight may be made so small in comparison with the weight of the suspended masses that the error introduced by it is almost negligible. We may however get rid of the error altogether by hanging a piece of the same cord from the bottoms of the two weights, forming a sort of tail-rope. The weight of cord on each side of the pulley will then always be the same and it will only be necessary to include the whole mass of the cord with the mass moved.

95. It is almost unnecessary to remark that the experiments detailed above have never been carried out precisely in the form there given, nor would the results of such experiments, were they attempted, accurately correspond with those stated, the principal cause of the discrepancies being the mass of the pulley C, for the experiments described were supposed to be conducted with an apparatus in which the mass of the pulley was insensible. They however serve to illustrate the method which might be pursued by a person ignorant of dynamics, and investigating the laws of motion and of gravitation by means of Attwood's machine, the results at which he would arrive agreeing sufficiently nearly with those stated above to suggest the *probable* truth of those laws which have been deduced from what would be the result of experiments performed under the ideal conditions which we have supposed to obtain. The evidence on which these laws are accepted must then be looked for from the agreement with observed phenomena of calculations based on the *assumption* of their truth. The experiments described also illustrate the methods in which such experiments may be varied so that the "question asked" in each may be in the simplest form possible, and the answer obtained may be simply the answer to *that question*,

and not involve the answers to several others from which it could be disentangled only by repeatedly varying the conditions.

96. It may be well here briefly to review the several units which have been already defined, and to show how each is connected with the three fundamental units of time, length, and mass. These three units are chosen arbitrarily, and upon them the magnitudes of all others employed in dynamical science depend.

The unit of time universally adopted throughout the world for scientific purposes is the second of mean solar time, and is therefore ultimately derived from observations of the earth's rotation. All measurements of velocities amount, therefore, simply to a comparison of the motion of the body considered with that of the earth about its axis, time being employed merely as the connecting link.

The unit of length adopted in Britain by engineers is the foot. This is the third part of the distance between the centres of two lines engraved on two gold plugs, sunk in a bar of bronze, which is now kept at the Exchequer Chambers, and known as the Imperial Standard Yard, the temperature of the bar at the time of observation being 62° Fahrenheit.

The unit of mass adopted in British measurements is the Imperial pound, that is, the quantity of matter contained in a certain mass of Platinum, kept at present in the Exchequer Chambers, and known as the Imperial Standard Pound Avoirdupois.

97. We pass on now to the consideration of the units derived from the three fundamental units of time, length, and mass.

The unit of velocity is the velocity of a point which passes over the unit of length in the unit of time.

If a second and a foot are the units of time and length, the unit of velocity is the velocity of a point which passes over one foot in a second.

Suppose that when the unit of time is τ seconds and the unit of length σ feet, the velocity of a certain point is denoted by u ; what will be the measure of this velocity when the unit of time is t seconds and the unit of length s feet ?

The number of feet in u units of length, each containing σ feet, is $u\sigma$. Let v be the measure required. Then

in τ seconds the point passes over $u\sigma$ feet ;

∴ in 1 second   ,,   ,, $\dfrac{u\sigma}{\tau}$ feet ;

∴ in t seconds   ,,   ,, $u\sigma\dfrac{t}{\tau}$ feet.

But $u\sigma\dfrac{t}{\tau}$ feet are equivalent to $u\dfrac{\sigma}{s}\dfrac{t}{\tau}$ units of length, each containing s feet; therefore

$$v = u \cdot \dfrac{\sigma}{s} \cdot \dfrac{t}{\tau};$$

that is, a velocity which is denoted by u, when τ seconds and σ feet are the units of time and length respectively, will be denoted by $u \cdot \dfrac{\sigma}{s} \cdot \dfrac{t}{\tau}$, when t seconds and s feet are units.

The numerical measure of a velocity, therefore, varies inversely as the unit of length and directly as the unit of time. This result is also a consequence of the fact, already stated, that the *unit* of velocity varies directly as the unit of length, and inversely as the unit of time. (See Arts. 8–10.)

98. The unit of acceleration is the acceleration of a point whose velocity is increased by the unit of velocity in the unit of time.

If a second and a foot are the units of time and length, the unit of acceleration is the acceleration of a point whose velocity is increased in one second by one foot per second.

Suppose a certain acceleration to be denoted by f *when* τ *seconds is the unit of time and* σ *feet the unit of length ; what will be the measure of this acceleration when* t *seconds and* s *feet are the units of time and length respectively ?*

MEASURE OF ACCELERATION. 65

The velocity generated in τ seconds with the given acceleration is a velocity of f units of length per τ seconds, that is, of $f\sigma$ feet per τ seconds, since the unit of length is σ feet. Hence,

with the given acceleration in τ'' there is generated a velocity per τ'' of $f\sigma$ feet;

$\therefore$ with the given acceleration in $1''$ there is generated a velocity per τ'' of $f\dfrac{\sigma}{\tau}$ feet;

$\therefore$ with the given acceleration in $1''$ there is generated a velocity per $1''$ of $f\dfrac{\sigma}{\tau^2}$ feet;

$\therefore$ with the given acceleration in $1''$ there is generated a velocity per t'' of $f\dfrac{\sigma t}{\tau^2}$ feet;

$\therefore$ with the given acceleration in t'' there is generated a velocity per t'' of $f\sigma \dfrac{t^2}{\tau^2}$ feet.

Now $f\sigma \dfrac{t^2}{\tau^2}$ feet are equivalent to $f\dfrac{\sigma}{s} \cdot \dfrac{t^2}{\tau^2}$ of the new units of length, each of which contains s feet. Hence with the given acceleration there is generated in the new unit of time a velocity such that if a point were moving with this velocity it would pass over a distance equal to $f\dfrac{\sigma}{s} \cdot \dfrac{t^2}{\tau^2}$ of the new units of length in the new unit of time. Hence $f\dfrac{\sigma}{s} \cdot \dfrac{t^2}{\tau^2}$ is the measure of the acceleration referred to the new units, that is, an acceleration which, when τ seconds and σ feet are the units of time and length, is denoted by f, will be denoted by $f\dfrac{\sigma}{s} \cdot \dfrac{t^2}{\tau^2}$, when t seconds is the unit of time and s feet the unit of length.

From this result it will be seen that the numerical measure of an acceleration varies inversely as the unit of length and directly as the square of the unit of time; a result previously found by considering the variation of the *unit* of acceleration. (See Arts. 13 and 14.)

G. D.   F

99. As examples we may take the following:

Ex. 1. *The acceleration produced by gravity in a particle falling freely being denoted by* 32, *when a second and a foot are the units of time and length, what will be the measure of this acceleration when a day and the length of the earth's radius are units, the latter being supposed equal to* 4000 *miles?*

Gravity in $1''$ generates a velocity per $1''$ of 32 feet;

∴ ,, ,, 1 day ,, ,, ,, $1''$,, $32 \times 24 \times 60^2$ ft.;

∴ ,, ,, 1 day ,, ,, 1 day of $32 \times 24^2 \times 60^4$ ft.

Now $32 \times 24^2 \times 60^4$ feet are equivalent to $\dfrac{32 \times 24^2 \times 60^4}{5280 \times 4000}$ times the earth's radius. The required measure is therefore $\dfrac{32 \times 24^2 \times 60^4}{5280 \times 4000}$ or $11310\tfrac{6}{11}$.

100. **Ex. 2.** *An acceleration which, when a second and a foot are units, is represented by* 32·2, *is represented by* 9660, *a yard being the unit of length. Find the unit of time.*

Let t seconds be the unit of time. Then with the acceleration in t'' there is generated a velocity per t'' of 9660 yds.

∴ ,, ,, ,, $1''$,, ,, t'' ,, $\dfrac{9660}{t}$,,

∴ ,, ,, ,, $1''$,, ,, $1''$,, $\dfrac{9660}{t^2}$,,

that is, of $\dfrac{9660 \times 3}{t^2}$ feet.

But with the acceleration in one second there is generated a velocity of 32·2 feet per second;

∴ $\dfrac{9660 \times 3}{t^2} = 32\cdot 2$;

∴ $t = 30$,

and the unit of time is thirty seconds, or half a minute.

101. The unit of momentum is the momentum possessed by the unit of mass when moving with the unit of velocity.

UNITS OF MOMENTUM AND FORCE. 67

The unit of momentum, therefore, varies directly as the unit of mass and directly as the unit of velocity: but this latter varies directly as the unit of length and inversely as the unit of time. Therefore the unit of momentum varies directly as the unit of mass, directly as the unit of length, and inversely as the unit of time. Hence, since the numerical measure of any quantity varies inversely as the unit in terms of which it is measured, it follows that the numerical measure of a given momentum varies inversely as the unit of mass, inversely as the unit of length, and directly as the unit of time.

If the measure of a momentum referred to any known set of units be given, its measure in terms of any other system of units can be found by the same method as that adopted above in the case of acceleration, provided that each of the second system of units is known in terms of the corresponding unit in the first set.

102. The unit of force is that force which generates the unit of momentum in the unit of time: or, that force which acting upon the unit of mass for the unit of time generates in it the unit of velocity.

If a second, a foot, and a pound be taken as the units of time, length, and mass respectively, the unit of force is that force which acting on the mass of a pound for a second generates in it a velocity of one foot per second. This force we have shown to be rather less than the weight of half an ounce.

A certain force is represented by P *when* τ *seconds,* σ *feet, and* μ *pounds are the units of time, length, and mass respectively: what will be the measure of this force when* t *seconds,* s *feet, and* m *pounds are the respective units?*

Since the force is represented by P in the first system of units, and a force is measured by the momentum which it will generate in the unit of time, it follows that the force acting on a mass of μ pounds for τ seconds will generate in it a velocity of P units of length, that is, of $P\sigma$ feet, per τ seconds.

∴ the force acting on μ lbs. for τ'' generates a velocity per τ'' of $P\sigma$ feet;

∴ the force acting on 1 lb. for τ'' generates a velocity per τ'' of $P\sigma\mu$ feet;

∴ the force acting on 1 lb. for $1''$ generates a velocity per τ'' of $P\sigma\mu\dfrac{1}{\tau}$ feet;

∴ the force acting on 1 lb. for $1''$ generates a velocity per $1''$ of $P\sigma\mu\dfrac{1}{\tau^2}$ feet;

∴ the force acting on 1 lb. for $1''$ generates a velocity per t'' of $P\sigma\mu\dfrac{t}{\tau^2}$ feet;

∴ the force acting on 1 lb. for t'' generates a velocity per t'' of $P\sigma\mu\dfrac{t^2}{\tau^2}$ feet;

∴ the force acting on m lbs. for t'' generates a velocity per t'' of $P\sigma\dfrac{\mu}{m}\cdot\dfrac{t^2}{\tau^2}$ feet.

But $P\sigma\dfrac{\mu}{m}\cdot\dfrac{t^2}{\tau^2}$ feet are equivalent to $P\dfrac{\sigma}{s}\cdot\dfrac{\mu}{m}\cdot\dfrac{t^2}{\tau^2}$ of the new units of length. Hence the force acting on the new unit of mass for the new unit of time will generate in it a velocity of $P\dfrac{\sigma}{s}\cdot\dfrac{\mu}{m}\cdot\dfrac{t^2}{\tau^2}$ new units of length per the new unit of time. The measure of the force, expressed in terms of the new system of units, is therefore $P\dfrac{\sigma}{s}\cdot\dfrac{\mu}{m}\cdot\dfrac{t^2}{\tau^2}$.

The numerical measure of a force, therefore, varies inversely as the unit of mass, inversely as the unit of length, and directly as the square of the unit of time. This also follows immediately from the fact that the unit of force varies directly as the unit of mass, directly as the unit of length, and inversely as the square of the unit of time.

103. *Ex. 1. Suppose that a force is represented by* 32·18 *when a second, a foot, and a pound are the units of time, length, and mass respectively: what will be the*

EXAMPLES OF THE CHANGE OF UNITS. 69

measure of the force when the unit of length is the centimetre, and the unit of mass the gramme, the unit of time remaining the same; it being given that a gramme is equal to 15·432...*grains, and a centimetre to* ·03281...*feet?*

These latter units are those employed in electrical and magnetic measurements, and the problem amounts to finding how many of the absolute units of force adopted by electricians are equivalent to the weight of an Imperial pound in London, since such weight is very nearly equal to 32·18 British absolute units of force.

The given force acting on 7000 grains for 1″ generates a velocity per 1″ of 32·18 feet;

∴ the given force acting on 1 grain for 1″ generates a velocity per 1″ of 32·18 × 7000 feet;

∴ the given force acting on 15·432 grains for 1′ generates a velocity per 1″ of $\dfrac{32\cdot18 \times 7000}{15\cdot432}$ feet.

Now a centimetre is equivalent to ·03281...feet; hence $\dfrac{32\cdot18 \times 7000}{15\cdot432}$ feet are equivalent to $\dfrac{32\cdot18 \times 7000}{15\cdot432 \times \cdot03281}$ centimetres. The given force will, therefore, in one second generate in a gramme a velocity of $\dfrac{32\cdot18 \times 7000}{15\cdot432 \times \cdot03281}$, or 444,893 centimetres per second, very nearly; hence 444,893 is the measure required.

From this we see that the weight in London of an Imperial pound of matter is approximately equal to 444,893 absolute units of force when a second is the unit of time, a centimetre the unit of length, and a gramme the unit of mass. The unit of force belonging to the centimetre-gramme-second system is called a *dyne*.

104. Ex. 2. *If the unit of force be the weight of the unit of mass, the unit of length being a foot, what must be the unit of time on the supposition that the acceleration produced by gravity in a body falling freely is denoted by* 32 *when a second and a foot are units of time and length respectively?*

The unit of force, acting on the unit of mass for the

unit of time, generates in it the unit of velocity. Let t seconds be the unit of time.

Then the weight of the unit of mass generates in that mass in t seconds a velocity of one foot per t seconds.

Therefore the weight of the unit of mass generates in that mass in one second a velocity of $\frac{1}{t}$ feet per t seconds, that is, a velocity of $\frac{1}{t^2}$ feet per second.

But the weight of any mass generates in that mass in one second a velocity of 32 feet per second.

Therefore $\quad \frac{1}{t^2} = 32$, or $t = \frac{1}{4\sqrt{2}}$.

Hence the unit of time required is $\frac{1}{4\sqrt{2}}$ seconds.

105. **Ex. 3.** *If the unit of force be the weight of 10 pounds, and the unit of acceleration when referred to a second and a foot as units be denoted by 8, find the unit of mass, assuming that the value of g when a second and a foot are units is 32.*

Let m pounds be the unit of mass. The unit of force is that force which acting upon the unit of mass produces in it the unit of acceleration.

Therefore the weight of 10 pounds acting on a mass of m pounds produces in it an acceleration which, when a second and a foot are units, is denoted by 8.

Therefore the weight of 1 pound acting on a mass of m pounds produces in it an acceleration which, when a second and a foot are units, is denoted by $\frac{8}{10}$.

Therefore the weight of 1 pound acting on a mass of 1 pound produces in it an acceleration which, when a second and a foot are units, is denoted by $\frac{8m}{10}$.

But the weight of 1 pound acting on a mass of 1 pound produces in it an acceleration which, when a second and a foot are units, is denoted by 32.

EXAMPLES OF THE CHANGE OF UNITS. 71

Therefore $\quad \dfrac{8m}{10} = 32$;

$\therefore m = 40$;

or the unit of mass is 40 pounds.

106. Ex. 4. *The unit of mass being a ton, the unit of force the weight of one hundred-weight, and the unit of velocity a velocity of 8 feet per second, it is required to find the units of time and length, assuming that the velocity generated in one second by gravity in a body falling freely is a velocity of 32 feet per second.*

Let t seconds be the unit of time.

The unit of force is that force which produces the unit of acceleration in the unit of mass.

Hence the unit of acceleration is that which is produced in the mass of a ton by a force equal to the weight of 1 hundred-weight.

Now the weight of 1 ton generates in 1 second in the mass of 1 ton a velocity of 32 feet per second.

Therefore the weight of 1 hundred-weight will generate in a second in the mass of 1 ton a velocity of $\dfrac{32}{20}$ feet per second.

The unit of acceleration, therefore, is that with which in one second there is generated a velocity of $\dfrac{32}{20}$ feet per second.

Therefore with the unit of acceleration there is generated in t seconds, that is, in the unit of time, a velocity of $\dfrac{32}{20}t$ feet per second.

But with the unit of acceleration there is generated in the unit of time the unit of velocity, which is a velocity of 8 feet per second.

Therefore $\quad \dfrac{32}{20}t = 8$;

$\therefore t = 5$.

Hence the unit of time is 5 seconds.

Again, the unit of velocity is a velocity of 8 feet per

second, that is, a velocity of 8×5, or 40, feet per unit of time.

But the unit of velocity is the velocity of a point which passes over the unit of length in the unit of time.

Therefore the unit of length is 40 feet.

107. The unit of impulse is the impulse of a force which generates the unit of momentum; or which, acting upon the unit of mass, generates in it the unit of velocity.

Suppose a given impulse to be denoted by I *when* τ *seconds,* σ *feet, and* μ *pounds are the units of time, length, and mass respectively: what will be the measure of this impulse when* t *seconds is the unit of time,* s *feet the unit of length, and* m *pounds the unit of mass?*

The given force will generate I units of velocity in the unit of mass, that is, a velocity of $I\sigma$ feet per τ seconds;

∴ the given force generates in μ pounds a velocity per τ'' of $I\sigma$ feet;

∴ the given force generates in 1 pound a velocity per τ'' of $I\sigma\mu$ feet;

∴ the given force generates in 1 pound a velocity per $1''$ of $I\sigma\mu\dfrac{1}{\tau}$ feet;

∴ the given force generates in 1 pound a velocity per t'' of $I\sigma\mu\dfrac{t}{\tau}$ feet.

∴ the given force generates in m pounds a velocity per t'' of $I\sigma\dfrac{\mu}{m}\cdot\dfrac{t}{\tau}$ feet.

But $I\sigma\dfrac{\mu}{m}\cdot\dfrac{t}{\tau}$ feet are equivalent to $I\dfrac{\sigma}{s}\cdot\dfrac{\mu}{m}\cdot\dfrac{t}{\tau}$ units of length each consisting of s feet.

Hence the given force will generate in the new unit of mass a velocity of $I\dfrac{\sigma}{s}\cdot\dfrac{\mu}{m}\cdot\dfrac{t}{\tau}$ new units of length per the new unit of time. $I\dfrac{\sigma}{s}\cdot\dfrac{\mu}{m}\cdot\dfrac{t}{\tau}$ is therefore the measure of the impulse of the force in terms of the new system of units.

UNIT OF WORK.

We see then that the numerical measure of an impulse varies inversely as the unit of mass, inversely as the unit of length, and directly as the unit of time. This might also have been deduced, as in the preceding cases, by considering the variation of the unit of impulse.

108. Ex. *Suppose an Armstrong shaft of 700 pounds, moving with a velocity of 1200 feet per second, to strike an iron target, and to be brought to rest: what will be the measure of the impulse when one hundred-weight is the unit of mass, a yard the unit of length, and a minute the unit of time?*

The force of impact, during its action, destroys in 700 pounds a velocity of 1200 feet per second;

∴ it would destroy

in 1 lb. a velocity per 1″ of 1200×700 feet.
„ 1 „ „ „ „ 60″ „ $1200 \times 700 \times 60$ „
„ 112 lbs. „ „ 60″ „ $\dfrac{1200 \times 700 \times 60}{112}$ „

Now $\dfrac{1200 \times 700 \times 60}{112}$ feet are equivalent to 150,000 yards.

Hence the impulsive force would destroy (or generate) in the unit of mass a velocity of 150,000 units of length per the unit of time.

Therefore 150,000 is the measure required.

109. The unit of work is the work done by the unit of force when its point of application moves through the unit of length in the direction of the force.

The variation in the measure of a given amount of work when the fundamental units are changed may be determined in precisely the same way as the variation in the measure of a given force or impulse. It will be found that the numerical measure of a given amount of work varies inversely as the unit of mass, inversely as the square of the unit of length, and directly as the square of the unit of time.

We will illustrate this by some examples.

Ex. 1. *The units of time, length, and mass being the*

74   EXAMPLES OF THE CHANGE OF UNITS.

second, centimetre, and gramme respectively, how many of the corresponding units of work are equivalent to the foot-pound; g being represented by $32 \cdot 18$ *when a second and a foot are units of time and length?*

We have seen in a previous example (see Art. 103), that, when g has the value given above, the weight of an Imperial pound is approximately equivalent to 444,893 of the absolute units of force belonging to the centimetre-gramme-second system of units.

Therefore the work done by a force equal to the weight of a pound when its point of application moves through a distance of one centimetre in the direction of the force is 444,893 centimetre-gramme-second units of work, or *ergs*.

But a centimetre is equal to $\cdot 03281$ feet.

Therefore the work done by a force equal to the weight of a pound when its point of application moves through a foot in the direction of the force is equivalent to $\dfrac{444893}{\cdot 03281}$ centimetre-gramme-second units of work: hence the number of these units of work contained in a foot-pound is $\dfrac{444893}{\cdot 03281}$, or 13,559,676, very nearly.

110. Ex. 2. *If an agent working at the rate of one horse-power perform the unit of work in the unit of time, and the acceleration produced by gravity in a body falling freely be the unit of acceleration, a pound being the unit of mass, find the units of time and length, it being given that g is equal to* 32 *when a second and a foot are units.*

Let t seconds be the unit of time.

The agent performs 33,000 foot-pounds of work per minute, or 550 foot-pounds per second.

Therefore in the unit of time, that is, in t seconds, it performs $550t$ foot-pounds.

But it performs the unit of work in the unit of time.

Therefore $550t$ foot-pounds is the unit of work.

Now the unit of force is that force which acting on the unit of mass produces in it the unit of acceleration.

EXAMPLES OF THE CHANGE OF UNITS. 75

Therefore the unit of force is that force which, acting on the mass of a pound, produces in it the same acceleration as is produced in it by gravity.

The unit of force is therefore equal to the weight of a pound.

Again, the unit of work is the work done by the unit of force when its point of application moves through a distance equal to the unit of length, and in the direction in which the force acts, and the unit of work has been shown to be $550t$ foot-pounds.

Therefore the unit of length is $550t$ feet.

Again, the unit of acceleration is the acceleration of a point in which in the unit of time there is generated a velocity with which if a point move it will pass over the unit of length in the unit of time.

∴ with the unit of acceleration

in t'' there is generated a velocity per t'' of $550t$ feet.
„ $1''$ „ „ „ „ „ t'' „ 550 „
„ $1''$ „ „ „ „ „ $1''$ „ $\dfrac{550}{t}$ „

But with the unit of acceleration in one second there is generated a velocity of 32 feet per second.

Therefore

$$\dfrac{550}{t} = 32;$$

$$\therefore t = \dfrac{550}{32} = 17\tfrac{3}{16}.$$

The unit of time is therefore $17\tfrac{3}{16}$ seconds.

Also the unit of length was shown to be $550t$ feet.

Therefore the unit of length is equal to

$550 \times 17\tfrac{3}{16}$ feet;
$= 9453\tfrac{1}{8}$ feet.

111. The unit of density is the density of a uniform substance, the unit of volume of which contains the unit of mass.

It will be seen that the unit of density, and therefore

the measure of the density of any substance, is independent of the unit of time, and depends only on the units of length and mass.

Also the unit of density must vary directly as the unit of mass if the unit of volume remain constant; for if the unit of mass be changed, the mass of the unit of volume of the substance whose density is unity will be changed in the same ratio. Again, if the unit of volume be changed, the unit of mass remaining the same, the volume which must contain the unit of mass of the substance whose density is unity will be changed in the same ratio: therefore, the standard substance must be changed so that the mass of a *given* volume of it may be changed in the inverse ratio of the unit of volume. Therefore, if the unit of mass remain constant, the density of the substance whose density is unity, that is, the unity of density, will vary inversely as the unit of volume. If the unit of volume be the volume of a cube whose edge is the unit of length, it follows that the unit of density varies inversely as the cube of the unit of length. Also it has been shown that the unit of density varies directly as the unit of mass when the unit of volume remains constant. Hence, when all are allowed to vary together, the unit of density will vary directly as the unit of mass, and inversely as the cube of the unit of length.

Since the numerical measure of any quantity varies inversely as the unit in terms of which it is measured, it follows that the numerical measure of the density of any substance varies inversely as the unit of mass, and directly as the cube of the unit of length.

112. *Suppose the density of a substance to be represented by ρ when μ pounds is the unit of mass, and σ feet the unit of length: what will be the measure of the same density when* m *pounds is the unit of mass and* s *feet the unit of length?*

The unit of volume of the substance contains ρ units of mass when μ pounds and σ feet are units of mass and length.

EXAMPLES OF THE CHANGE OF UNITS. 77

Therefore σ^3 cubic feet of the substance contain $\mu\rho$ pounds.

∴ 1 cubic foot ,, ,, contains $\dfrac{\mu\rho}{\sigma^3}$,,

∴ s^3 cubic feet ,, ,, contain $\mu\rho\dfrac{s^3}{\sigma^3}$,,

But $\rho\mu\,\dfrac{s^3}{\sigma^3}$ pounds are equivalent to $\rho\,\dfrac{\mu}{m}.\dfrac{s^3}{\sigma^3}$ units of mass each containing m pounds.

Therefore the new unit of volume of the substance contains $\rho\,\dfrac{\mu}{m}.\dfrac{s^3}{\sigma^3}$ new units of mass.

The measure of the density in terms of the new system of units is therefore $\rho\,\dfrac{\mu}{m}.\dfrac{s^3}{\sigma^3}$.

113. **Ex.** *If the unit of force be the weight of one ounce, and the mass of a cubic foot of the substance whose density is unity be* 162 *pounds, the unit of time being one second, what is the unit of length, g being equal to* 32 *when a second and a foot are units?*

Let the unit of length be s feet.

Then the unit of volume is s^3 cubic feet.

Since a cubic foot of the standard substance contains 162 pounds of matter, the unit of volume of this substance must contain $162s^3$ pounds.

But the mass of the unit of volume of the standard substance is the unit of mass.

Therefore the unit of mass is $162s^3$ pounds.

Now the unit of force is that force which acting on the unit of mass for the unit of time generates a velocity of the unit of length per the unit of time.

Therefore, the weight of

	oz.				lbs.			feet.
1 generates in 1″	in a mass of	$162s^3$	a velocity per 1″ of	s
1 lb. ,, ,, ,, ,,	$162s^3$,, ,, ,,	$s \times 16$
1 ,, ,, ,, ,, ,,	1 lb. ,,	$s \times 16 \times 162s^3$.

But the weight of one pound acting on the mass of one

pound for a second generates a velocity of 32 feet per second.

Therefore $162s^4 \times 16 = 32$;

$$\therefore s = \frac{1}{3}.$$

The unit of length is therefore one-third of a foot, that is, 4 inches.

APPENDIX ON THE DIMENSIONS OF UNITS.

114. A LINE, whether straight or curved, being "length without breadth," possesses only one degree of extension in space. In the neighbourhood of any point, not being a singular point on a curve, the line extends in only one direction and its opposite, and is the simplest form of extension of which we can conceive. A line is therefore said to be of *one dimension* in space, a point being that which has no dimensions, but position only.

The magnitude of a line is completely known when we know its length, and is expressed in terms of the unit of length. Thus, if we say that the length of a straight line is l, we mean that it contains l units of length, and if $[L]$ denote the unit of length, the complete expression for the length of the line is $l\,[L]$. Now if the unit of length $[L]$ vary while the line considered remains invariable, the expression $l\,[L]$ must remain constant while $[L]$ varies, and therefore l must vary inversely as $[L]$; in other words, the numerical measure of a given line varies *inversely* as the unit of length in terms of which the line is measured, and therefore *inversely* as the number expressing the unit of length in terms of some invariable unit.

115. A superficies, whether plane or curved, possesses "only length and breadth," and is therefore extended in two directions. A superficies is consequently said to be of *two dimensions* in space, and its magnitude is completely known when we know its area. All areas are measured in terms of the unit of area which is the unit of two dimensions in space. Thus when we say that the area of a superficies is a, we mean that the figure con-

UNIT OF AREA. 79

tains a units of area, and if $[A]$ denote the unit of area, the complete expression for the magnitude of the superficies is $a\,[A]$, and, as in the case of a line, we see that if $[A]$ varies while the figure considered remains unchanged, a must vary inversely as $[A]$, or the numerical measure of an area varies *inversely* as the unit of area.

Now, invariably in mathematical investigations, and generally in practical measurements, the unit of area adopted is the area of a square the length of whose side is the unit of length: and this may be represented by $[L]^2$. It must be borne in mind that the symbol $[L]^2$ does not represent an Algebraical product, since $[L]$ is not a number, but it represents a *geometrical quantity of a totally different nature from that expressed by* $[L]$, which is defined as the area of a square whose side is equal to the quantity $[L]$.

116. It may be proved geometrically that the number of units of area in any rectangle, when the unit of area is defined as above, is equal to the product of the numbers of units of length in two adjacent sides, and therefore the number of units of area in any square is equal to the algebraical square of the number of units of length in its side. The areas of different squares are therefore to each other in the duplicate ratio of the number of units of length in their respective sides, and therefore *different units of area are to one another in the ratio of the squares of the numbers expressing the measures of the corresponding units of length in terms of one and the same unit.* The words placed in italics are very important; from them it follows, since the numerical measure of an area varies inversely as the unit of area, that *the numerical measure of an area varies inversely as the square of the number expressing the unit of length in terms of some invariable unit.* For the sake of brevity it is frequently stated that the numerical measure of an area varies inversely as the square of the unit of length, but the two latter terms of the proportion must be reduced to numerical measures before they can be Algebraically compared. With this understanding we shall in future generally use the abbreviated form of expression.

UNIT OF VOLUME AND MASS.

117. A solid is extended in three directions in space, since it possesses length, breadth, and thickness, and is said to be of *three dimensions* in space. Its magnitude is known when we know its volume, and is expressed in terms of the unit of volume. Thus when we say that the volume of a solid is c, we mean that it contains c units of volume, and denoting the unit of volume by $[C]$ the complete expression for this is $c\,[C]$, and as in the other cases considered, the numerical measure of a given volume varies *inversely* as the unit of volume.

118. The unit of volume generally adopted is the volume of a cube, whose edge is equal to the unit of length, and this may be represented by $[L]^3$, this symbol representing a physical quantity of a totally different nature from those considered in the preceding articles, and not capable of being measured in terms of them. By reasoning similar to that of the preceding articles it may be shown that, when the unit of length varies, the volume of this cube varies directly as the cube of the number expressing the unit of length in terms of some constant unit, and we therefore infer that the numerical measure of a volume varies *inversely as the cube of the number expressing the unit of length in terms of a constant unit*, or, as it is generally expressed, inversely as the cube of the unit of length.

119. We have thus seen that the numerical measures of quantities of one, two, and three dimensions in space vary *inversely* as the first, second, and third powers of the unit of length respectively. We shall not proceed to consider higher dimensions in space because most persons are unable to form a distinct conception of space of more than three dimensions, but shall now turn our attention to physical quantities of a different nature.

120. The magnitude or duration of an interval of time is measured in terms of the unit of time, and if we say that its measure is t we mean that it consists of t units of time, and the complete expression for the interval is $t[T]$, where $[T]$ denotes the unit of time. The numerical measure of a given interval will therefore vary *inversely* as the unit of time in terms of which it is measured. By

analogy an interval of time may be said to be of *one dimension* in time. At present we know of nothing of more than one dimension in time, though we shall presently meet with quantities of negative dimensions in time.

121. The quantity of matter in a body, that is, its mass, is measured in terms of the unit of mass, and if its numerical measure be m, the complete expression for the mass of the body is $m\,[M]$, where $[M]$ represents the unit of mass. The numerical measure of a quantity of matter varies *inversely* as the unit of mass, while mass, like time, being a primary conception, its unit is completely arbitrary. A quantity of matter may be said to be of *one dimension* in mass.

122. Having thus considered the primary units we proceed to the consideration of complex or derived units; that is, of units derived from the three fundamental units of time, length, and mass.

The unit of velocity is the velocity of a point which passes over the unit of space in the unit of time. If v denote the measure of the velocity of a point and $[V]$ the unit of velocity, the complete expression for the velocity will be $v\,[V]$, where v represents a number only, and $[V]$ a physical quantity. Now the space passed over in the time t by a point moving with uniform velocity v is equal to vt units of length, and is therefore completely represented by $vt\,[L]$, or by $l\,[L]$, where l is a number equal to vt. But the complete representation of the product of the velocity $v\,[V]$ in the time $t\,[T]$ is $v\,[V]\,t\,[T]$. Hence we have the equation

$$vt\,[V]\,[T] = vt\,[L],$$
or
$$[V]\,[T] = [L].$$
Therefore
$$[V] = \frac{[L]}{[T]} \quad \ldots\ldots\ldots\ldots\ldots\ldots\ldots(\text{I}),$$
$$= [L]\,[T]^{-1} \ldots\ldots\ldots\ldots\ldots(\text{II}),$$

or the unit of velocity is of the nature of the line of unit length divided by the unit of time. This is expressed by saying that the unit of velocity is of one dimension in length, and of minus one dimension in time, as shown by equation (II).

The unit of velocity therefore varies *directly* as the unit of length, and *inversely* as the unit of time, and since the numerical measure of a given velocity varies inversely as the unit of velocity, the numerical measure of a velocity varies *inversely* as the unit of length, and *directly* as the unit of time.

123. The unit of acceleration is that of a point whose velocity is changed by the unit of velocity in the unit of time. If $[F]$ denote the unit of acceleration, the complete representation of an acceleration whose measure is f is $f[F]$, and f varies inversely as $[F]$ if the acceleration remain constant. Also the velocity generated in t units of time is ft units of velocity, and the complete representation of the product when a point moves with an acceleration $f[F]$ for a time $t[T]$ is the product $f[F] t[T]$. Hence we have the equation

$$ft [F] [T] = ft [V];$$

therefore $$[F] = \frac{[V]}{[T]},$$

or an acceleration may be said to be of one dimension in velocity and of minus one dimension in time. But by the preceding article

therefore $$[V] = [L] [T]^{-1};$$
$$[F] = [L] [T]^{-2}.$$

The unit of acceleration is therefore of *one dimension* in length, and of *minus two dimensions* in time, and varies *directly* as the unit of length, and *inversely as the square* of the unit of time. Hence, since the measure of any acceleration varies inversely as the unit of acceleration, it follows that the numerical measure of a given acceleration varies *inversely* as the unit of length and *directly* as the square of the unit of time.

124. We may remark that whenever in the definition of a quantity the word *per* is introduced, it implies that the quantity is of minus one dimension in the element immediately succeeding the *per*, and if the word *per* is

UNITS OF MOMENTUM AND FORCE.

introduced twice in the definition, each time it is introduced it implies minus one dimension in the quantity immediately succeeding. Thus velocity is space passed over *per* unit of time, and is of minus one dimension in time, while the unit of acceleration is that which generates *per* unit of time a velocity of unit length *per* unit of time, and is of minus two dimensions in time. Similarly, if the price of oranges be so much *per* dozen oranges, the price is of minus one dimension in oranges, and the product of the price into a number of oranges is an amount of money, showing that the price is also of one dimension in money, and the measure of the price will vary directly as the number of oranges *per* which it is quoted.

125. The unit of momentum is the momentum possessed by the unit of mass moving with the unit of velocity. If $[K]$ represent the unit of momentum, a momentum whose measure is k will be represented by $k\,[K]$, and if this momentum remain constant k must vary inversely as $[K]$. Also the momentum of a mass m moving with velocity v is mv units of momentum, that is $mv\,[K]$. But the momentum of a particle is possessed by it in virtue of the two factors, its mass and velocity, and is completely represented by their product, that is, by $m\,[M]\,v\,[V]$; hence

$$mv\,[K] = mv\,[M][V],$$
therefore $\quad [K] = [M]\,[V]$
$\qquad\qquad = [M]\,[L]\,[T]^{-1}.$

The unit of momentum is therefore of one dimension in mass, one in length, and minus one in time, and the measure of a given momentum will therefore vary *inversely* as the unit of mass, *inversely* as the unit of length, and *directly* as the unit of time.

126. The unit of force is that force which generates the unit of momentum in the unit of time, and the product of the numerical measure of a force into the number of units of time during which it acts is the number of units of momentum generated thereby. If p be the numerical measure of a force, and $[P]$ represent the unit of force, the complete representation of the force is $p\,[P]$, and

while the force remains constant p varies inversely as $[P]$. But the *effect* of a force p acting for a time t is to generate pt units of momentum, and the complete representation of the product of a force p $[P]$ during a time t $[T]$ is p $[P]$ t $[T]$. Hence

Therefore
$$pt\ [P]\ [T] = pt\ [K].$$
$$[P]\ [T] = [K].$$
$$= [M]\ [L]\ [T]^{-1},$$
and
$$[P] = [M]\ [L]\ [T]^{-2}.$$

Hence the unit of force is of one dimension in mass, one in length, and minus two in time, and the measure of a given force will therefore vary *inversely* as the unit of mass, *inversely* as the unit of length, and *directly* as the *square* of the unit of time.

These conclusions might have been deduced from the consideration that force is that which produces acceleration in mass, and is measured by the product of the mass and of the acceleration produced therein.

127. The unit of work $[W]$ is that which is performed by the unit of force when its point of application moves over the unit of length in the direction of the force. The work done by any force measured in terms of this unit of work is equal to the product of the measure of the force and the number of units of length passed over by its point of application in the direction of the force. Hence if the point of application of a force p $[P]$ move over a space l $[L]$ in the direction of the force, the work done is completely represented by pl $[W]$. But the product of a force p $[P]$ working through a distance l $[L]$ is completely represented by p $[P]$ l $[L]$. Hence

or
$$pl\ [W] = pl\ [P]\ [L],$$
$$[W] = [P]\ [L]$$
$$= [M]\ [L]^2\ [T]^{-2}.$$

The unit of work is therefore of one dimension in mass, two in length, and minus two in time, and the numerical measure of a given amount of work will therefore vary *inversely* as the unit of mass, *inversely* as the *square* of the unit of length, and *directly* as the *square* of the unit of time.

UNITS OF ENERGY AND DENSITY.

128. The unit of kinetic energy $[E]$ is *twice* the energy possessed by a particle moving with unit momentum and unit velocity, or, in other words, by a particle of unit mass moving with the unit of velocity. The kinetic energy of any moving particle is measured in terms of this unit by one-half the product of its momentum and velocity. Hence if its momentum be k $[K]$ and its velocity v $[V]$, its energy is $\frac{1}{2} kv$ $[E]$. But one-half the product of the momentum and velocity is represented completely by

$$\frac{1}{2} k [K] v [V].$$

Hence $\quad \frac{1}{2} kv [E] = \frac{1}{2} kv [K] [V],$

therefore $\quad [E] = [K] [V] = [M] [V]^2$
$\qquad\qquad = [M] [L]^2 [T]^{-2}.$

The unit of energy is therefore of one dimension in mass, two in space, and minus two in time. It is therefore of precisely the same nature as the unit of work, and we shall see (Art. 146) that when work is done on a free particle, an *equivalent* amount of energy is *always* given to it; this could not be the case were work and energy of different natures, that is, of different dimensions in the three fundamental mechanical units. The numerical measure of the kinetic energy of a moving particle will, of course, vary *inversely* as the unit of mass, *inversely* as the square of the unit of length, and *directly* as the square of the unit of time.

129. The unit of density $[D]$ is the density of a body the unit of volume of which contains the unit of mass. The density of any homogeneous body is measured in terms of this unit by the number of units of mass *per* unit of volume, and we therefore infer that while the unit of density is of one dimension in mass it is of *minus* one dimension in *volume*, that is, of minus three dimensions in length, since volume is of three dimensions in space. The same conclusion might have been derived from the consideration that the mass of any homogeneous

body is the product of its volume and density, or, if $v'\,[L]^3$ be its volume, $d\,[D]$ its density, its mass is $v'd\,[M]$, whence, as in other cases, we get

$$[D]\,[L]^3 = [M],$$

or
$$[D] = \frac{[M]}{[L]^3}.$$

The numerical measure of the density of a given body will therefore vary *inversely* as the unit of mass, and *directly* as the cube of the unit of length.

130. If the body considered be of the nature of a string, then its linear density, that is, its mass *per unit of length*, is of one dimension in mass, and of minus *one* in length, as shown by the *per*, and its measure will vary directly as the unit of length, and inversely as the unit of mass.

If the body be a thin lamina, its surface density, that is, its mass *per unit of area*, is of one dimension in mass and of minus one in area, that is, of minus two in space, and the numerical measure of its density will therefore vary *inversely* as the unit of mass, and *directly* as the *square* of the unit of length.

131. The unit of impulse is the impulse of a force which generates the unit of momentum. An impulse is measured by the number of units of momentum generated. Hence, if $[I]$ denote the unit of impulse, and i the measure of a given impulse, the momentum corresponding to the impulse $i\,[I]$ is i units of momentum, and this is denoted by $i\,[K]$, where $[K]$ represents the unit of momentum, and we have therefore

$$i\,[I] = i\,[K],$$

or
$$[I] = [K]$$
$$= [M]\,[L]\,[T]^{-1}. \quad \text{(Art. 125.)}$$

The unit of impulse is therefore of one dimension in mass, one in length, and minus one in time, and the numerical measure of an impulse varies *inversely* as the unit of mass, *inversely* as the unit of length, and *directly* as the unit of time.

ASTRONOMICAL UNIT OF MASS.

Impulse is of the same dimensions and therefore of the same nature as momentum, just as kinetic energy is of the same dimensions and nature as work.

132. As another example, we will consider the dimensions of the *Astronomical* unit of mass in terms of the fundamental units. The Astronomical unit of mass is defined as that quantity of matter which acting upon an equal quantity of matter at unit distance (each mass being supposed condensed at a point), attracts it with the unit of force. The law of gravitation is, that the attraction between two material particles varies directly as the product of their masses, and inversely as the square of the distance between them. Hence, if m_1, m_2 denote the masses of two particles expressed in terms of the Astronomical unit, and r the measure of the distance between them, the number of units of force with which each attracts the other is $\frac{m_1 m_2}{r^2}$. Hence, if $[N]$ denote the unit of mass, we have

$$\frac{m_1 [N] \, m_2 [N]}{r^2 [L]^2} = \frac{m_1 m_2}{r^2} [P],$$

where $[P]$ represents the unit of force.

But the unit of force being that force which acting on the unit of mass produces in it the unit of acceleration, is of one dimension in mass, one in length, and minus two in time, and is therefore represented by $[N][L][T]^{-2}$ (Art. 126),

or $\qquad [P] = [N][L][T]^{-2}$,

therefore $\dfrac{m_1 [N] \, m_2 [N]}{r^2 [L]^2} = \dfrac{m_1 m_2}{r^2} [N][L][T]^{-2}$,

therefore $\qquad \dfrac{[N]}{[L]^2} = [L][T]^{-2}$,

therefore $\qquad [N] = [L]^3 [T]^{-2}$.

The Astronomical unit of mass is therefore of *three* dimensions in space and of *minus two* in time.

The unit of density is of one dimension in mass and of minus three in length. Hence the Astronomical unit of density is of *minus two* dimensions in time only. The Astronomical unit of density therefore depends solely on the unit of time and is independent of the unit of length.

133. Each of the above units might, of course, be expressed in terms of any of the others instead of the fundamental units; for example, the unit of mass might be said to be of one dimension in force, minus one in length, and of two in time; the unit of acceleration, of one dimension in work or energy, minus one in length, and minus one in mass; the unit of time, of one dimension in velocity, and of minus one in acceleration, and so on; but it is very seldom indeed that such expressions are of service. It is highly important to remember the dimensions of all the units in terms of the fundamental units of mass, length, and time, since then we can immediately find the measure of any quantity in terms of any given system of units.

134. *For example, suppose it required to find how many units of work of the metric system are equivalent to a foot-pound, the unit of mass being the gramme, that is,* 15·432 *grains, and the unit of length the centimetre, that is,* ·03281 *feet the value of g in British measure being* 32·1912.

The unit of work is of one dimension in mass, two in length, and minus two in time. Hence the measure of a quantity of work varies inversely as the unit of mass, inversely as the square of the unit of length, and directly as the square of the unit of time. In the present example the unit of time remains unaltered. The foot-pound is equal to 32·1912 British absolute units of work. Hence the measure of a foot-pound, referred to the above metric units, is

$$32 \cdot 1912 \times \frac{7000}{15 \cdot 432} \times \left(\frac{1}{\cdot 03281}\right)^2,$$

$$= 13{,}564{,}400 \text{ nearly.}$$

135. The units recommended for general use by the Committee of the British Association are those belonging to the centimetre-gramme-second system, in which the unit of length is the centimetre or ·03281 feet, and the unit of mass the gramme, or 15·432 grains, the unit of time being one second.

BRITISH ASSOCIATION UNITS. 89

The c.g.s. *unit of force* is that force which acting on a gramme of matter for a second generates in it a velocity of one centimetre per second, and is called a *dyne*.

The c.g.s. unit of work is the work done by *the dyne in working through a centimetre*, and is called an *erg*, or *ergon*. The *unit of energy* being mechanically equivalent to the unit of work the same name is applied to it.

The c.g.s. *unit of power* is the power of doing work at the rate of *one erg per second*.

A rate of doing work equal to 10^7 ergs per second is called a *watt*.

For multiplication or division by a million the prefixes *mega-*, or *megal-*, and *micro-*, may be employed; thus, a *megadyne* is a million of dynes and a *microdyne* the millionth of a dyne.

The prefixes *kilo-*, *hecto-*, *deca-*, *deci-*, *centi-*, *milli-*, may also be employed in their usual senses, that is to say, the prefixes kilo-, hecto-, deca-, deci-, centi-, milli-, imply respectively multiplication by $1000, 100, 10, \frac{1}{10}, \frac{1}{100}, \frac{1}{1000}$, so that a *kilo-gramme* is a thousand grammes while a *centigramme* is the one-hundredth of a gramme.

For the expression of high decimal multiples and submultiples the exponent of the power of ten which serves as multiplier is denoted, when positive, by an appended cardinal number, and when negative, by a prefixed ordinal number.

Thus 10^9 grammes constitute a *gramme-nine*, while $\frac{1}{10^9}$ of a gramme constitutes a *ninth-gramme*. A megalerg is equivalent to an *erg-six*, while a microdyne is a *sixth-dyne*.

The weight of a *gramme* is about 980 *dynes* or rather less than a *kilodyne*.

The weight of a *kilogramme* is rather less than a *megadyne*.

The dyne is about 1·02 of the weight of a milligramme

at any point of the earth's surface; and the megadyne about 1·02 of the weight of a kilogramme.

The *kilogrammetre*, or work done against gravity in lifting a kilogramme one metre in height, is rather less than the *ergon-eight*, being about 98,000,000 *ergs*.

The *gramme-centimetre* is rather less than the *kilerg*, being about 980 ergs.

The value of g in the above statements is taken as 980 c.g.s. units of acceleration.

The *weight of a pound* is about 445,048 *dynes*.

A *foot-pound* is equivalent to about 13,564,400 *ergs*, or rather more than 13½ *megalergs*.

One *horse-power* is about 7·46 *erg-nines* per second, or 746 *watts*.

Nearly the whole of this article has been taken almost verbatim from the report of the British Association for the Advancement of Science for 1873.

136. The dimensions of the principal mechanical units have already been given, but for the sake of convenience they are tabulated below.

TABLE SHOWING THE DIMENSIONS OF UNITS.

Length = $[L]$, Mass = $[M]$, Time = $[T]$,
Area .... = $[L]^2$,
Volume .... = $[L]^3$,
Density .... = $[M][L]^{-3}$,
Velocity................................. = $[L][T]^{-1}$,
Acceleration .... = $[L][T]^{-2}$,
Momentum or Impulse .... = $[M][L][T]^{-1}$,
Force .... = $[M][L][T]^{-2}$,
Work or Energy .... = $[M][L]^2[T]^{-2}$,
Power or Rate of Doing Work ... = $[M][L]^2[T]^{-3}$.

137. Having determined the dimensions of all the quantities with which we have to deal in terms of the fundamental units, it is easy to translate any set of measurements from one system to another. The process is in fact little else than a process of construing with the assistance of the table of dimensions as a vocabulary.

When it is required to find the fundamental units on which any system of measurements is founded, it is frequently advisable to assume symbols for the ratios of the units to the foot, pound and second, and then to translate all the measurements into the foot-pound-second system. Illustrations of the process will be found in the examples of Arts. 138–142.

It should be noticed that the numerical measure of the kinetic energy of a particle is *one-half* the product of the momentum and velocity. Hence the kinetic energy of unit mass moving with unit velocity is only half a unit, and the unit of kinetic energy is that possessed by two units of mass moving with unit velocity or by unit mass moving with $\sqrt{2}$ units of velocity. The unit of work, which corresponds to the unit of kinetic energy is that done by the unit of force in working through unit distance. Now acting on the unit of mass the unit of force will produce unit of acceleration, and will therefore in unit time generate unit velocity. The energy of the unit of mass moving with this velocity is only half a unit of energy, so that, acting on a unit of mass originally at rest, the unit of force in unit time produces only half a unit of energy, and in the next chapter we shall see that during this time the body is moved through only half the unit of length, so that the force does only half a unit of work.

137a. Power is rate of doing work. Now the work done by an agent is the product of the force exerted into the distance through which its point of application is moved in its own direction. Hence the rate at which an agent works is the product of the force exerted and the velocity of the point of application in the direction of the force. Thus the tractive force of a locomotive expressed in pounds' weight, multiplied by the speed at which it is running expressed in feet per second, gives its rate of working in foot-pounds per second, and dividing by 550 we obtain its rate of working in *horse-power*. Similarly, if a rope is employed for traction, the tension of the rope multiplied by its velocity gives the power transmitted.

137β. When a belt or strap passing over a pulley or

drum is employed to drive a shaft the power transmitted may be determined in the same manner, but in this case the belt pulls both at the top and bottom of the pulley, and it is the difference of the tensions in the advancing and receding portions of the belt which we have to consider. For suppose the belt is equally tight throughout. Then the pulley is solicited in opposite directions by the two portions of the belt acting upon its circumference with equal forces, and if the pulley meet with no resistance and be already in motion, it will continue to move, but no power will be transmitted. If the motion of the pulley were resisted, its speed could only be maintained by the receding portion of the belt exerting a greater pull upon it than the approaching portion. Suppose that T_1 is the tension in the part of the belt which is leaving the pulley, T_2 the tension in that which is approaching the pulley, and suppose the velocity of the belt to be v feet per second. Then in one second the work done by the receding part of the belt is vT_1, but of this vT_2 units are employed in overcoming the pull of the approaching portion or slack part of the belt; hence the work done per second on the pulley is $v(T_1 - T_2)$ which is therefore the measure of the power transmitted. By expressing v in feet per second, T_1 and T_2 in pounds' weight, and dividing the expression $v(T_1 - T_2)$ by 550, we obtain the horse-power transmitted.

The maximum ratio of T_1 to T_2 is determined by the condition that the belt shall not slip, and as machinery is generally subject to vibration a certain margin of safety is required in determining this ratio. It will depend on the nature of the pulley and of the belt and on the angle of contact. If the "driver" and "follower" are of the same diameter the angle of contact will be about 180°. If the driver and follower are nearly in the same horizontal plane the angle of contact may be made somewhat greater than 180° by making the lower portion of the belt the tight portion and the upper the slack portion, especially if the pulleys are at a considerable distance apart, for the "sag" of the slack portion, being greater than that of the tight portion, increases the angle of contact. Under such circumstances with wrought or cast

EXAMPLES OF CHANGE OF UNITS. 93

iron pulleys and clean leather belts T_1 may with safety be taken to be about three times T'_2.

137γ. When energy is transmitted along a shaft it is the couple exerted and the angular velocity of the shaft that determines the power transmitted. For suppose the moment of the couple exerted on the shaft to be G foot-pounds and that the shaft turns through the angle a in one second. Then we may suppose the couple due to $\frac{G}{r}$ pounds' weight acting at the extremity of an arm of length r feet. In turning through an angle a the extremity of the arm would move through ar feet and the work done by the force would be $\frac{G}{r} ar$ foot-pounds, that is Ga foot-pounds, a being, of course, expressed in radians (circular measure). Hence if a shaft be exposed to a couple of G foot-pounds and turn through the angle a in a second, the power transmitted is Ga foot-pounds per second. If the shaft make n revolutions per second $a = 2n\pi$ and the power transmitted is $2n\pi G$ foot-pounds per second. The horse-power transmitted is obtained by dividing this expression by 550.

137δ. For example:—*A dynamo absorbs* 25 H.P. *and is driven by a belt running at* 100 *feet per second, the tension in the tight portion being* $2\frac{1}{2}$ *times that in the slack portion of the belt. Find the tensions.*

Hence the difference of the tensions is $\frac{3}{2}T_2$. The power absorbed is 550×25 foot-pounds per second and the velocity of the belt is 100 feet per second.

Hence $\frac{3}{2}T_2 \times 100 = 550 \times 25$.

$$\therefore T_2 = \frac{275}{3} = 91\tfrac{2}{3}.$$

and $$T_1 = \frac{5}{2}T_2 = 229\tfrac{1}{6}.$$

The tension of the tight portion is therefore equal to the weight of $229\tfrac{1}{6}$ pounds; that of the slack portion to the weight of $91\tfrac{2}{3}$ pounds.

138. The following are examples of units:—

Ex. I. *If* 10,000 *foot-pounds be the unit of work, the weight of* 5 *cwt. the unit of force and* 1 *cwt. the unit of mass, find the units of length and time.*

Let L feet be the unit of length and T seconds the unit of time.

The dimensions of force are $\frac{[M][L]}{[T]^2}$ and those of work are $\frac{[M][L]^2}{[T]^2}$.

Hence the unit of force will be equal to $\frac{112.L}{T^2}$ foot-pound-second units and the unit of work to $\frac{112.L^2}{T^2}$ foot-pound-second units.

But the weight of 5 cwt. is 560×32 poundals and 10,000 foot-pounds is 320,000 foot-poundals.

$$\therefore \frac{112L}{T^2} = 560 \times 32,$$

and $$\frac{112L^2}{T^2} = 320{,}000,$$

$$\therefore L = 17\tfrac{6}{7} \text{ and } T = \frac{5}{4\sqrt{14}}.$$

139. Ex. II. *If the unit of velocity be that of a point which passes over* 8 *feet in* 3 *seconds and the unit of acceleration be that in virtue of which a velocity of* 60 *miles an hour is generated in* 55 *seconds, find the units of space and time.*

Let L feet be the unit of length, and T seconds the unit of time.

The unit of velocity will be $\frac{L}{T}$ feet per second, and the unit of acceleration will be $\frac{L}{T^2}$ foot-second units.

Now a velocity of 8 feet in 3 seconds is $\frac{8}{3}$ feet per second; 60 miles an hour is 88 feet per second, and the

EXAMPLES ON THE CHANGE OF UNITS. 95

acceleration in virtue of which this velocity is produced in 55 seconds is $\frac{88}{55}$ or $\frac{8}{5}$ foot-second units.

$$\therefore \frac{L}{T} = \frac{8}{3} \text{ and } \frac{L}{T^2} = \frac{8}{5},$$

$$\therefore T = \frac{5}{3} \text{ and } L = \frac{40}{9}.$$

140. **Ex. III.** *In the first of a system of units the measure of a certain acceleration is 500 times its measure in the second system, but the measure of a certain velocity is only 50 times its measure in the second system. Find the ratios of the units of length and time.*

Let L_1 feet and T_1 seconds be the units in the first system; L_2 feet and T_2 seconds the units in the second system. Suppose that the acceleration is f foot-second units and the velocity v feet per second.

In the first system the acceleration will be represented by $f \frac{T_1^2}{L_1}$ and the velocity by $v \frac{T_1}{L_1}$. In the second system they will be represented by $f \frac{T_2^2}{L_2}$ and $v \frac{T_2}{L_2}$ respectively.

$$\therefore f \frac{T_1^2}{L_1} = 500 f \frac{T_2^2}{L_2}$$

and $$v \frac{T_1}{L_1} = 50 v \frac{T_2}{L_2}.$$

Hence $T_1 = 10\ T_2$ and $L_1 = \frac{1}{5} L_2,$

or $\frac{T_1}{T_2} = 10$ and $\frac{L_1}{L_2} = \frac{1}{5}.$

141. **Ex. IV.** *If the density of wrought iron be denoted by 12, and the work done in lifting 1 cwt. vertically through 20 feet by 14, a pound being the unit of mass, find the units of length and time, it being given that 1 cub. foot of wrought iron contains 480 lbs.*

Let L feet be the unit of length and T seconds the unit of time. The unity of density is therefore that of a substance of which L^3 cub. ft. contain 1 lb. and the unit of work is $\dfrac{L^2}{T^2}$ foot-poundals.

The density of wrought iron will therefore be represented by $480L^3$ and the work done in lifting 1 cwt. through 20 feet by $\dfrac{112 \times 20 \times 32 T^2}{L^2}$.

$$\therefore 480L^3 = 12,$$

and $$\dfrac{112 \times 20 \times 32 T^2}{L^2} = 14,$$

$$\therefore L = \dfrac{1}{2\sqrt[3]{5}} \text{ and } T = \dfrac{1}{64 \cdot 5};$$

142. **Ex. V.** *In a certain system of absolute units the acceleration of gravity is denoted by* 10, *the kinetic energy of a train of* 200 *tons travelling at* 60 *miles an hour by* 40, *and its momentum by* 2. *Find the units of length, mass, and time.*

Let L feet be the unit of length, M pounds the unit of mass, and T seconds the unit of time.

The acceleration of gravity is 32 foot-second units, and will be denoted by $32\dfrac{T^2}{L}$.

The energy of the train is $\dfrac{200 \times 2240 \times 88^2}{2}$ foot-poundals, and will be denoted by $\dfrac{200 \times 2240 \times 88^2}{2} \cdot \dfrac{T^2}{ML^2}$.

Its momentum is $200 \times 2240 \times 88$ foot-second-pound units, and will be denoted by $200 \times 2240 \times 88 \dfrac{T}{ML}$.

$$\therefore 32 \dfrac{T^2}{L} = 10 \ldots\ldots\ldots\ldots\ldots\ldots(1),$$

$$\dfrac{200 \times 2240 \times 88^2}{2} \cdot \dfrac{T^2}{ML^2} = 40 \ldots\ldots\ldots\ldots(2),$$

$$200 \times 2240 \times 88 \cdot \dfrac{T}{ML} = 2 \ldots\ldots\ldots\ldots\ldots(3).$$

From (2) and (3)

$$44\frac{T}{L} = 20 \quad\quad\quad (4),$$

But
$$32\frac{T^2}{L} = 10 \quad\quad\quad (1),$$

$$\therefore \frac{8}{11}T = \frac{1}{2} \text{ or } T = \frac{11}{16},$$

$$\therefore \text{ by (4) } L = \frac{121}{80},$$

and
$$M = 200 \times 2240 \times 44\frac{T}{L}$$
$$= 200 \times 2240 \times 20 \text{ by (4)}$$
$$= 8,960,000.$$

EXAMINATION ON CHAPTER I.

NOTE. In all the examples, except where otherwise stated, the numerical value of g referred to a foot and a second as units of length and time is taken to be 32.

1. How must a physical quantity be measured? Of what does the complete representation of any physical quantity consist?

2. Define the velocity of a point.

If a second be the unit of time and an acre be represented by 10, what will be the measure of a velocity of 45 miles an hour?

3. If a train move from rest with uniform acceleration, and in five minutes attain a velocity of 60 miles per hour, find the measure of its acceleration when a second is the unit of time and a foot the unit of length.

4. What must we know about an acceleration in order that it may be completely defined? Show that an acceleration can at any time be represented by a straight line.

5. A ship is sailing due North with a velocity of 10 knots an hour, while another is steaming South-West at the rate of 15 knots an hour. Find the velocity of the second ship relative to the first.

6. A ship whose head points N.N.E. is steaming at the rate of 16 knots an hour in a current which flows E.S.E. at the rate of 4 knots an hour, find the velocity of the ship relative to the sea-bottom.

7. Explain what is meant by the resultant of two independent accelerations, and what by the acceleration of one point relative to another.

8. Define the density of a body. If one pound be the unit of mass and a yard the unit of length, find the measure of the density of water, it being given that a cubic foot of water contains 1,000 ozs.

9. State Newton's second law of motion, and explain briefly the nature of the evidence on which our acceptance of this and other physical laws is based.

10. What do you understand by the physical independence of forces?

11. What is the dynamical unit of force?

If the unit of mass be a ton, the unit of length a yard, and the unit of time a minute, compare the unit of force with the weight of one pound, taking g equal to 32, when a foot and second are units.

12. During what time must a constant force equal to the weight of one ton act upon a train of 100 tons to generate in it a velocity of 40 miles per hour?

13. Upon what experimental evidence do we base the assertion that the attraction of the earth upon any body is proportional to its mass and independent of the nature of the material of which it is formed?

14. Supposing the attraction of gravitation at the equator to be ·995 of its value in London, if a person sell goods in London by a spring balance accurately graduated at the equator, how much per cent. on the selling price does he gain in excess of his fair profit?

15. If the cage of a lift be descending with an acceleration represented by $\frac{1}{10} g$, find the pressure which a man of 12 stone exerts upon the bottom of the cage.

Account for the fact that a person after descending the shaft of a coal-pit in the cage, when being brought to rest near the bottom, feels as if he were being lifted up again.

16. Explain fully what is meant by the equation $P = mf$.

If the unit of length be a yard, find the unit of time in order that the weight of a body may be numerically equal to its mass.

17. When is a force said to do work?

If the force required to draw a carriage on a level road be equal to the weight of $37\frac{1}{2}$ pounds, how many absolute units of work does a horse do in drawing the same carriage from Ely to Cambridge, a distance of 17 miles?

If the horse take two hours in performing the journey, compare the rate at which he works with a horse-power.

18. If the unit of velocity be a velocity of 45 miles per hour, and the unit of acceleration when referred to a foot and a second as units of length and time be represented by 11, find the units of length and time.

19. If the weight of one pound be the unit of force, a velocity of eight feet per second the unit of velocity, and the acceleration produced by gravity in a particle falling freely be one-third of the unit of acceleration, find the units of mass, length, and time.

20. If the density of water be the unit of density, and 10 lbs. the unit of mass, find the unit of length, it being given that a cubic foot of water contains 1,000 ozs.

21. The density of water being the unit of density, the weight of a cubic foot of water the unit of force, and a pound the unit of mass, find the units of length and time.

22. If the unit of momentum be that possessed by a mass of 10 lbs. after falling freely from rest during one second, and the unit of kinetic energy be that possessed by a pound after falling freely from rest for two seconds, find the unit of mass and the unit of velocity.

23. If 33,000 foot-pounds be the unit of work, the

weight of a ton the unit of force, and 5 cwt. the unit of mass, find the units of length and time.

24. If an engine perform 100 units of work in the unit of time when the unit of mass is a hundred-weight, the unit of acceleration that produced by gravity in a particle falling freely, and the unit of velocity a velocity of 100 feet per second, compare the rate at which the agent works with a horse-power.

EXAMPLES ON CHAPTER I.

1. Supposing the earth to rotate about its axis in 23 hours 56 minutes, its equatorial diameter being 7925 miles, find the velocity of a point at the equator relative to the earth's centre in feet per second, and in miles per minute.

2. What is the measure of a velocity of 45 miles an hour (1) in feet per second, (2) in chains per minute?

3. If a day were the unit of time and a thousand miles the unit of length, what would be the numerical measure of the velocity of the earth's centre about the sun, supposing it to describe a circle of 92,000,000 miles radius uniformly in $365\frac{1}{4}$ days?

4. A ship is sailing due North at the rate of 12 knots an hour, and another is steaming due East at 16 knots an hour, find the velocity of the first relative to the second, the ships being so near together that the surface of the water may be considered plane.

5. Two straight railway lines make an angle of 60° with each other, and two trains are running each at the rate of 40 miles an hour away from the point of intersection of the lines, one on one line and one on the other. Find the direction and magnitude of their relative velocity.

6. A passenger in a railway carriage observes another train moving on a parallel line in the opposite direction to occupy two seconds in passing him, but if the other

train had been proceeding in the same direction as the observer, it would have appeared to pass him in 30 seconds. Compare the rates of the two trains.

7. A force which can statically support 50 lbs. acts uniformly for one minute on a mass of 200 lbs.; find the velocity and momentum acquired by the body.

8. The mass of a balloon and its appendages is 2 tons, and that of the air displaced by it is 4800 lbs. Find the acceleration with which it will begin to ascend.

9. The mass of a train is 200 tons, the resistance arising from friction, etc., 12 lbs. weight per ton. If the tractive force upon it be equal to the weight of a ton and a half, find its acceleration.

10. A mass of 1000 tons initially at rest is acted on by a force constant in magnitude and direction, and equal to the weight of 14 lbs. After what interval will it have a velocity of 1 foot per second?

11. A force equal to the weight of 300 lbs. acts constantly at an inclination of 30° to a line AB. If its point of application move 100 feet parallel to the line in 1 minute, find the work done by the force, and the rate of work in horse-power.

12. Show that the work done in lifting weights to different heights from the same horizontal plane is equal to the work done in lifting the sum of the weights to a height equal to the height of their centre of gravity in their final positions.

A shaft 10 feet in diameter has to be sunk to a depth of 130 fathoms, through chalk: how much work must be expended in raising the materials if the mass of a cubic foot of chalk be 2315 ozs.?

13. Show that the work done in dragging a body up a rough inclined plane is the same as that which would be done in dragging it horizontally along a distance equal to the base of the plane, the coefficient of friction being the same as in the first case, and in lifting it vertically through a height equal to the height of the plane.

14. If a particle slide down a rough inclined plane, show that its acceleration is $g \dfrac{\sin(\alpha - \phi)}{\cos \phi}$ where $\tan \phi$ is equal to the coefficient of friction.

15. If the mass of the Scotch express be 150 tons, and the resistances to its motion arising from the air, friction, etc., amount to 16 lbs. weight per ton, when the train is going at the rate of 60 miles an hour on a level plain, find the horse-power of the engine which can just keep it going at that rate.

16. If, in the preceding example, the driving-wheels of the engine be 8 feet 2 inches in diameter, and during each revolution of the wheels two pistons make each a complete stroke (to and fro), the diameter of each piston being 18 inches, and the length of the stroke 28 inches, find the mean *effective* pressure of the steam on each square inch of the pistons necessary to drive the train at 60 miles an hour, the slip of the driving wheels on the metals and the friction of the working parts of the engine being neglected.

Note. The steam must do as much *work* per revolution of the driving wheel on the pistons as is required to drive the train.

17. Find the actual horse-power of an engine which can just propel an ironclad ship at the rate of 16 knots an hour; the resistance to the ship's motion when steaming at that rate being equal to the weight of 50 tons, and a knot being taken equal to 6078 feet.

18. If a point situated at the orthocentre of a triangle have three component velocities, represented in magnitude and direction by its distances from the angular points of the triangle, show that its resultant velocity will tend to the centre of the circle circumscribing the triangle, and will be represented by twice the distance of the point from the centre.

19. A train travels at the rate of 45 miles an hour; rain is falling vertically, but owing to the motion of the train the drops appear, as they fall past the windows, to

EXAMPLES. 103

make an angle $\tan^{-1} 1\cdot 5$ with the vertical. Find the velocity of the raindrops.

20. If in Attwood's machine the string can bear a tension equal to only one-fourth the sum of the weights, show that the least acceleration possible is $\dfrac{g}{\sqrt{2}}$.

21. A shot of mass m is fired from a gun of mass M with a velocity u relative to the gun: show that, if the mass of the powder be neglected, the velocity of the shot is $\dfrac{Mu}{m+M}$, and that of the gun $\dfrac{mu}{m+M}$ relative to the ground.

22. A smooth wedge, whose angle is a, has one face in contact with a horizontal plane. Find the acceleration with which it must be made to move that a heavy particle may be in relative equilibrium on its inclined surface.

23. If a be the distance between two moving points at any time, V their relative velocity, and u, v the resolved parts of V in, and perpendicular to, the direction of a, show that their distance when they are nearest to each other is $\dfrac{av}{V}$, and that the time of arriving at this nearest distance is $\dfrac{au}{V^2}$.

24. If the acceleration caused by gravity be the unit of acceleration, and the velocity of a mile in 5 minutes the unit of velocity, find the unit of length.

25. If the units of length and time be a yard and a minute respectively, and the unit of force the weight of 32 lbs., find the unit of mass.

26. If the unit of time be 5 minutes, and the unit of length 5 yards, find the value of g.

27. If the acceleration of a falling body be the unit of acceleration, and a velocity of 3 miles an hour the unit of velocity, find the units of space and time.

28. If the unit of velocity be the velocity of a point which passes over a feet in t seconds, and the unit of

acceleration that of a point which acquires in τ seconds a velocity of b feet per τ seconds, find the units of length and time.

29. Two nations estimate the acceleration of gravity by numbers in the ratio of 300 to 1, but the velocity of the earth in space by numbers in the ratio of 5 to 1. Find the ratios of their units of time and length.

30. If the area of a ten-acre field be represented by 100, and the acceleration of a heavy falling particle by $58\frac{2}{3}$, find the unit of time.

31. If the unit of force be equal to the weight of 5 lbs. and the unit of acceleration, when referred to a foot and a second, be denoted by 3, find the unit of mass.

32. A constant force acts upon a particle during 3 seconds from rest, and then ceases; in the next 3 seconds it is found that the particle describes 180 feet: find the velocity of the particle at the end of the second second of its motion, and the numerical value of its acceleration (1) when a second, (2) when a minute, is taken as unit of time, the unit of length being 1 foot.

33. If the unit of velocity be a velocity of a feet per t seconds, and if the weight of 1 pound be the unit of force, and a pound the unit of mass, find the units of length and time.

34. If f_1, f_2 be the measures of an acceleration when $m+n$ seconds and $m-n$ seconds are the respective units of time, and a feet and b feet the respective units of length, show that the measure becomes $\frac{1}{c}(\sqrt{f_1 a} - \sqrt{f_2 b})^2$ when $2n$ seconds are taken as unit of time and c feet as unit of length.

35. The measures of an acceleration and a velocity, when referred to $(a+b)$ feet $(m+n)$ seconds, and $(a-b)$ feet $(m-n)$ seconds respectively, are in the inverse ratio of their measures when referred to $(a-b)$ feet $(m-n)$ seconds, and $(a+b)$ feet $(m+n)$ seconds. Their measures,

when referred to a feet m seconds, and b feet n seconds, are as $ma : nb$; show that

$$\frac{n}{m} = \sqrt{1 - \frac{b^4}{a^4}}.$$

36. If f be the measure of an acceleration when a feet and t seconds are units of space and time, and f' its measure when a' feet and t' seconds are units, and if the acceleration be measured by $f + f'$ when c feet and τ seconds are units, show that

$$\frac{\tau^2}{c} = \frac{t^2}{a} + \frac{t'^2}{a'}.$$

37. A shot of 700 lbs. is moving with a velocity of 1200 feet per second; find the numerical measure of its kinetic energy when a ton is the unit of mass, a yard the unit of length, and a minute the unit of time.

38. If the unit of length be a yard, the unit of acceleration that produced by gravity in a body falling freely, and the unit of density that of water, find the number of units of work required to raise a ton of ore from the bottom of a mine 600 fathoms deep, assuming that a cubic foot of water contains 1000 ozs.

39. Supposing the earth to be a sphere of 4000 miles radius and mean density 5·5 times that of water; if its mass be represented by a billion, the density of water by 10, and the work done in lifting a ton a yard high be the unit of work, find the units of mass, length, and time, a cubic foot of water containing 1000 ozs.

40. A circle revolves with uniform velocity in its own plane about its centre. The centre moves with varying velocity along a straight line. Find the velocity parallel to this line, at any instant, of a point on the circumference, and deduce the acceleration of the centre necessary for this point to be always moving at right angles to the line.

41. The number expressing the *weight* of a cubic foot of water (in absolute units) is $\frac{1}{10}$th of that expressing its volume, $\frac{1}{8}$th of that expressing its mass, and $\frac{1}{100}$th of the

number expressing the *work* done in lifting it one foot. Find the units of length, mass, and time.

42. The resistance to the motion of a train is equal to the weight of 12 lbs. per ton of its mass. An engine in drawing a train of 300 tons along a level line at 50 miles per hour consumes 20 lbs. of coal per minute. If the heat required to raise the temperature of one pound of water through 1° C. be capable of doing 1390 foot-pounds of work, and if the combustion of 1 lb. of coal be capable of raising 80 lbs. of water from the freezing to the boiling point, find how much of the whole heat generated is usefully employed by the engine.

43. Given that a quadrant of the earth's circumference is 10^9 centimetres, and that the mean density of the earth is 5·67, prove that the unit of force will be the attraction of two spheres each of 3928 grammes, whose centres are a centimetre apart; the acceleration of gravity at the earth's surface being 981, when a centimetre, second, and gramme are the units of length, time, and mass.

(N.B.—Two spheres attract as if they were particles of the same mass situated at their centres.)

44. In a certain system of absolute units the acceleration produced by gravity in a body falling freely is represented by 3, the kinetic energy of a 600 lb. shot, moving with a velocity of 1600 feet per second, is denoted by 100 and its momentum by 10; find the units of length, mass, and time, assuming that g is equal to 32 in the foot-pound-second system.

45. It is required to transmit 50 H.P. by a single leather belt, running at 100 feet per second, the tension of the tight portion being three times that of the slack. Find the width of belt necessary if the working tension be limited to 75 lbs. weight for a belt one inch wide.

CHAPTER II.

ON UNIFORM, AND UNIFORMLY ACCELERATED, MOTION.

143. IF a point move with uniform velocity v, the space passed over in t units of time will be denoted by vt. For the space passed over in each unit of time is v units of length, and therefore the space passed over in t units of time will be vt units of length.

If a material particle be under the action of no external force, it will remain at rest or move *uniformly* in a straight line. Hence, if such a particle be moving at any instant with velocity v, it will retain that velocity, and the space passed over in t units of time will be vt.

144. If a point move with a constant acceleration f in the direction of motion, the velocity generated in each unit of time is f units of velocity, and therefore the velocity generated in t units of time will be ft units of velocity. Hence, if the initial velocity of the point be u, the velocity, v, at the end of t units of time will be $u + ft$, and if the point be initially at rest, its velocity at the end of t units of time will be ft, or $v = ft$.

If the acceleration be in the direction opposite to that of the initial velocity, we must give the negative sign to f, and the above formula will still be true.

If a material particle of mass m be acted on by a constant force P in the direction of its motion, the momentum generated by the force at any instant will be in that direction, and the rate at which the momentum is generated will be constant, being proportional to the force. Hence the acceleration of the particle will always be in the direction of its motion and be constant. If f represents this acceleration, we have $mf = P$, the unit of force

108   UNIFORMLY ACCELERATED MOTION.

being properly chosen (see Art. 45). Hence if u be the initial velocity of the particle, the velocity at the end of t units of time will be $u + ft$, or $u + \dfrac{P}{m} t$, and its momentum will be $mu + Pt$.

145. To find the space passed over in t units of time by a point moving from rest with uniform acceleration f we may proceed as follows. Divide the time t into n equal intervals, each equal to τ. Then the velocities of the point at the beginning of the 1st, 2nd, 3rd,...n^{th} intervals will be denoted by 0, $f\tau$, $2f\tau$...$(n-1)f\tau$ respectively, since its acceleration is uniform. Also the velocities of the point at the end of the 1st, 2nd, 3rd...n^{th} intervals will be $f\tau$, $2f\tau$, $3f\tau$,...$nf\tau$ respectively. Hence, if the point moved during each interval with the velocity which it had at the beginning of the interval, the space passed over in the t units of time would be

$$0 \cdot \tau + f\tau \cdot \tau + 2f\tau \cdot \tau + \ldots + (n-1)f\tau \cdot \tau$$

$$= \frac{n(n-1)}{2} f\tau^2$$

$$= \frac{1}{2} fn^2\tau^2 \left(1 - \frac{1}{n}\right)$$

$$= \frac{1}{2} ft^2 \left(1 - \frac{1}{n}\right).$$

Again, if the point moved during each interval, with the velocity which it had at the end of the interval, the space passed over in the t units of time would be

$$f\tau \cdot \tau + 2f\tau \cdot \tau + 3f\tau \cdot + \ldots + nf\tau \cdot \tau$$

$$= \frac{n(n+1)}{2} f\tau^2$$

$$= \frac{1}{2} fn^2\tau^2 \left(1 + \frac{1}{n}\right).$$

Now the velocity of the point at any instant of a given interval is intermediate between its velocities at the beginning and end of that interval. Hence the space passed over during any interval is intermediate between that which would be passed over by the point if its velocity

KINETIC ENERGY EQUIVALENT TO WORK DONE. 109

remained constant during the interval, and the same as at the beginning of the interval, and that which would be passed over by the point if its velocity were the same throughout the interval as at the end of the interval. Hence the space actually passed over by the point during the t units of time will be intermediate between that which it would pass over under the first of the above hypotheses and that which it would pass over if the second hypothesis were realised, that is, the space actually passed over by the point is intermediate between

$$\frac{1}{2}ft^2\left(1-\frac{1}{n}\right) \text{ and } \frac{1}{2}ft^2\left(1+\frac{1}{n}\right),$$

and this is true however great n may be. Now as we diminish each interval and consequently increase n, each of the above hypotheses more nearly corresponds with what actually takes place; and if we make each interval indefinitely small, and therefore n indefinitely great, $\frac{1}{n}$ vanishes, and the above expressions for the space passed over ultimately coincide with each other, and therefore with the expression for the space actually passed over by the point, which is intermediate between the two. Hence the space passed over in t units of time by a point moving from rest with an acceleration f in the direction of motion is $\frac{1}{2}ft^2$ units of length.

146. From this we see that if a material particle, of mass m, move from rest under the action of a constant force P, the space passed over in t units of time will be $\frac{P}{2m}t^2$. The work done upon the particle will therefore (Art. 62) be $P \cdot \frac{P}{2m} \cdot t^2$. Its velocity will be $\frac{P}{m}t$ and its momentum Pt. Its kinetic energy will therefore (Art. 38) be $\frac{1}{2}Pt \cdot \frac{P}{m}t$, that is $\frac{P^2}{2m}t^2$, and is therefore numerically equal to the work done upon the particle by the force.

If at the end of the t units of time we suppose the

direction of the force reversed so as to retard the motion, the intensity of the force remaining the same as before; the velocity of the particle will be destroyed in the same time as that in which it was generated, and the velocity at any instant during the retardation will be the same as at the corresponding instant when the acceleration was positive. Hence the space passed over by the particle, while the velocity is being destroyed by the force P, is the same as that passed over during the production of that velocity. Hence the work done *against* the force *by* the particle while being brought to rest is the same as that done *by* the force *upon* the particle when the velocity of the latter was being increased, and is therefore numerically equal to the kinetic energy of the particle. The kinetic energy of a particle is therefore numerically equal to the amount of work which it is capable of doing in being brought to rest, and we thus see a reason for the term "energy" being applied to one half the product of the momentum and velocity of a moving particle.

147. The effect of a force when its point of application moves in its direction is to do work, and we now see that if it act upon a particle its effect is to generate an amount of kinetic energy numerically equal to the work done. Now these effects must be one and the same, and therefore kinetic energy is *mechanically* equivalent to work.

The effect of a given force, when its point of application moves over a given space in the direction of the force, is always to generate the same amount of kinetic energy represented by the Arithmetical product of these two factors; hence while the effect of a force acting for an interval of time is to generate an amount of momentum measured by the Algebraical product of *these* two factors, the effect of a force when its point of application is moved in its direction is to generate an amount of kinetic energy measured by the Arithmetical product of *these* factors. The distinction between these two products of a force is very important. If a given force act for a given time upon a particle, the distance through which it will move it will vary inversely as the mass, and therefore the kinetic energy generated will vary inversely as the mass

UNIFORMLY ACCELERATED MOTION. 111

while the momentum generated is invariable, but if the distance through which the particle is moved, instead of the time, remain the same, the kinetic energy generated will remain invariable, but the momentum produced will be proportional to the square root of the mass.

148. If a point move under the influence of a constant acceleration f in the direction of motion, but start with an initial velocity u, a process precisely similar to that adopted in Art. 145 will enable us to find the space passed over in any given time.

For, as before, dividing the time t into n equal intervals each equal to τ, the velocity of the point at the beginning of the respective intervals will be

$$u,\ u+f\tau,\ u+2f\tau,\ \ldots\ u+\overline{n-1}\,f\tau,$$

and its velocity at the end of these intervals will be

$$u+f\tau,\ u+2f\tau,\ u+3f\tau,\ \ldots\ u+nf\tau$$

respectively. Hence the space actually passed over in t units of time will be intermediate between

$$u\cdot\tau+(u+f\tau)\,\tau+(u+2f\tau)\,\tau+\ldots+(u+\overline{n-1}\cdot f\tau)\,\tau$$

and

$$(u+f\tau)\,\tau+(u+2f\tau)\,\tau+(u+3f\tau)\,\tau+\ldots+(u+nf\tau)\,\tau\ ;$$

that is, between

$$ut+\frac{1}{2}ft^2\left(1-\frac{1}{n}\right) \text{ and } ut+\frac{1}{2}ft^2\left(1+\frac{1}{n}\right);$$

and, increasing n indefinitely, each of these becomes ultimately equal to $ut+\frac{1}{2}ft^2$, which is therefore the space passed over by the point in t units of time.

149. The space passed over by a point moving under the influence of a constant acceleration in the direction of motion may also be found in the following way.

Let time be represented by lengths measured along the line AB, and let AB represent t units of time and contain t units of length. Let the velocity of the point at any time be represented by a straight line drawn perpendicular to AB from that point in AB which corresponds to the time in question, the line containing as many units

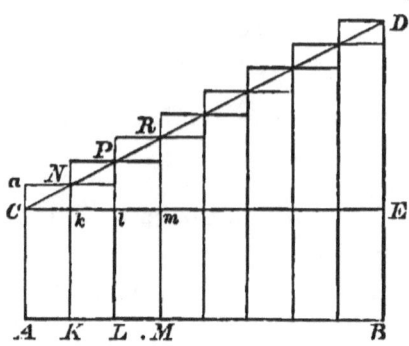

of length as the point possesses units of velocity. Let the time t be divided into n equal intervals, and let the straight line AB be divided into corresponding portions at the points K, L, etc. Through A, K, L, ... B, let lines AC, KN, LP, ... BD, be drawn perpendicular to AB, and representing the velocity of the moving point at the corresponding times. Draw the straight line Ckl ... E parallel to AB. Then, if u be the initial velocity of the point, and f its constant acceleration, $AC = u$, $KN = u + \dfrac{ft}{n} = u + f \cdot AK$, $LP = u + \dfrac{2ft}{n} = u + f \cdot AL$, and so on. Therefore $kN = f \cdot Ck$, $lP = f \cdot Cl$, and so for the other points. Hence C, N, P, ... D all lie in a straight line. Draw the straight line CD passing through each of the points N, P, etc. Complete the inner and outer series of parallelograms as in the figure. Then, if the moving point were to move during each interval with the velocity which it has at the beginning of the interval, the space passed over during any interval represented by LM will, since LP represents the velocity during that interval, contain as many units of length as the parallelogram PM contains units of area; and, this being true for each of the other intervals, it follows that the number of units of length which would be passed over by the point in the time represented by AB, that is, in t units of time, if it moved during each interval with the velocity which it actually has at the beginning of the interval, would be equal to the number of units of area contained in the sum of the inscribed parallelograms. Similarly the space passed over by the point in the same time, if it moved during each interval with the velocity which it actually has at the end of the interval, would be represented by the sum of the areas of the outer series of parallelograms. And this is true, however great may be the number of intervals into which the time t is divided. But the actual space passed over by the moving point must be intermediate between these two; and when the number of intervals into which the time is divided is indefinitely increased, the sums of the areas of the inner and outer series of parallelograms ultimately coincide with the area of the figure $CABD$. Hence the number of units of length passed over by the point in the time represented by AB is equal to the number of units of area in $CABD$. But the figure $CABD$ is made up of the rectangle $CABE$ and the triangle CED. Therefore its area is equal to

$$CA \cdot AB + \frac{1}{2} \cdot CE \cdot ED.$$

Now CA represents the initial velocity and therefore contains u units of length, DB represents the velocity at the end of time t and therefore contains $u + ft$ units of length; hence DE contains ft units of length, and AB or CE contains t units of length. Hence the area $CABD$ contains $ut + \frac{1}{2} \cdot ft \cdot t$, or $ut + \frac{1}{2} ft^2$ units of area. Therefore the space passed over in t units of time by a point starting with initial velocity u, and moving with a constant acceleration, f, in the direction of motion, contains $ut + \frac{1}{2} ft^2$ units of length.

If the point start from rest, $u = 0$, and the figure will contain no rectangle corresponding to $CABE$. The space passed over in the time represented by AB will then be represented by the area of the triangle CED, and will therefore be $\frac{1}{2} ft^2$ units of length.

Since DE contains ft units of length, and CE contains t

units, the ratio $\frac{DE}{EC}$ is numerically equal to f, and therefore the acceleration is represented geometrically in the figure by the tangent of the angle DCE.

(Compare Newton, Lemma x.)

150. If a material particle be under the action of a constant force in the direction of its motion, its acceleration will be constant, and the above investigation determines the space passed over by the particle in any given time, substituting for f the expression $\frac{P}{m}$, when m is the mass of the particle and P the force acting upon it.

As an example we may take the following.

A particle is allowed to fall from rest under the action of gravity only; find the space moved through by the particle in 4 seconds, supposing $g = 32$ when a foot and a second are units.

Here the only force acting on the particle is its weight, which is constant and equal to mg. Hence it will move with uniform acceleration g. The space passed over in 4 seconds will therefore be $\frac{1}{2}g \cdot 4^2$ ft. $= 8g$ ft. $= 256$ ft.

151. If we wish to find the space passed over in any particular second, the nth for example, by a point moving with uniform acceleration, we may find the space passed over in n seconds and subtract from it that passed over in $n-1$ seconds; or we may find the velocity at the beginning of the nth second, and then find the space passed over in 1 second by a point starting with this initial velocity and moving under the given acceleration.

For example, let it be required to find the space passed over by a particle falling freely, during the 7th second of its fall, g being supposed equal to 32.

The space passed over in 7 seconds by a particle falling from rest is $\frac{1}{2}g \cdot 7^2$ ft. $= 16 \times 7^2$ ft. That passed over in 6 seconds is 16×6^2 ft. The difference, or 16×13 ft., is the space passed over during the 7th second.

EQUATIONS FOR UNIFORMLY ACCELERATED MOTION. 115

Or we may proceed thus. The velocity at the end of the 6th second is 32×6 ft. Hence the particle at the beginning of the 7th second has an initial velocity of 32×6 ft. per second, and will therefore during that second pass over a distance equal to $\left\{(32 \times 6) + \dfrac{1}{2}g\right\}$ ft., or $32 \times 6\tfrac{1}{2}$ ft., putting $t = 1$ in the formula $ut + \dfrac{1}{2}ft^2$.

152. If a point move from rest with a constant acceleration f in the direction of motion, and if v represent its velocity at the end of t units of time, and s the space passed over by the point during that time, then we have

$$v = ft \quad \ldots\ldots\ldots\ldots\ldots\ldots(1),$$

$$s = \tfrac{1}{2}ft^2 \quad \ldots\ldots\ldots\ldots\ldots(2).$$

From (1) and (2) we get

$$v^2 = 2fs \quad \ldots\ldots\ldots\ldots\ldots(3).$$

These three equations are very important, and should be remembered.

If it be a material particle of mass m that is in motion, we obtain by multiplying each side of equation (3) by m and dividing by 2.

$$\tfrac{1}{2}mv^2 = mf \cdot s.$$

Now mf is the force which must act on the particle of mass m to produce the acceleration f; and since s is the space moved through by the particle in the direction of the force, $mf \cdot s$ is the work done upon the particle by the force, and we thus see that the kinetic energy of the particle is equivalent to the whole amount of work that has been done upon it since the commencement of the motion.

153. If a point start with velocity u, and move with a constant acceleration f in the direction of motion, and if v

represent its velocity at the end of t units of time, and s the space passed over by it during that time, we have

$$v = u + ft \quad\ldots\ldots\ldots\ldots(1),$$

$$s = ut + \tfrac{1}{2}ft^2 \quad\ldots\ldots\ldots(2).$$

Squaring (1) and multiplying (2) by $2f$, we see that

$$v^2 = u^2 + 2fs \quad\ldots\ldots\ldots(3).$$

If in this case it be a material particle of mass m that is in motion, then, multiplying each side of equation (3) by m, dividing by 2 and transposing, we get

$$\tfrac{1}{2}mv^2 - \tfrac{1}{2}mu^2 = mf \cdot s.$$

Here as before $mf \cdot s$ represents the work done upon the particle by the force producing acceleration, and the expression on the left-hand side of the equation represents the increase of the kinetic energy of the particle. Hence the increase of the kinetic energy of the particle during any time is equivalent to the amount of work done upon it during that time by the force producing acceleration.

If the direction of the acceleration of the moving point be opposite to that of its initial velocity, we have only to write $-f$ for f in the above expressions, and the equations will still be true.

154. It must be borne in mind that in all the above investigations the units of force, work, etc., are the absolute units belonging to the system of fundamental units employed. Thus if we adopt the foot-pound-second system the unit of force is the poundal which is $\tfrac{1}{g}$ of the weight of a pound; the unit of work is a foot-poundal or $\tfrac{1}{g}$ of a foot-pound, and the unit of energy is equivalent to the unit of work and is therefore also a foot-poundal.

Hence since the kinetic energy of a particle of mass m moving with velocity v is $\frac{1}{2}mv^2$ foot-poundals it is equal to $\frac{1}{2g}mv^2$ foot-pounds.

155. For example, the number of foot-pounds of work which can be done by a 600 lb. shot moving with a velocity of 1200 feet per second in coming to rest is
$$\frac{600 \times 1200^2}{2g} = \frac{600 \times 1200^2}{64} = 600 \times 150^2 = 13500000$$
foot-pounds.

Again, if in the course of one minute a steam gun project one hundred 4 lb. shots with a velocity of 1200 feet per second the work it does in one minute is
$$100\frac{4 \times 1200^2}{2g} = 9000000$$
foot-pounds, and the horse-power at which the gun works is
$$\frac{9000000}{33000} = 272\frac{8}{11}.$$

156. As illustrations of this portion of the subject we may take the following Examples.

Ex. 1. *A heavy particle is projected vertically upwards with velocity* u : *supposing its weight to be the only force acting upon it, it is required to completely determine the motion.*

The particle will in this case move with a constant acceleration g downwards. We must therefore write $-g$ for f in the equations of the preceding Article. Its velocity, v, at the end of the time t will therefore be given by
$$v = u - gt \dots\dots\dots\dots\dots\dots\text{(i)}.$$
The space s passed over in t units of time will be given by
$$s = ut - \frac{1}{2}gt^2 \dots\dots\dots\dots\dots\text{(ii)}.$$

If we take t equal to $\dfrac{u}{g}$, we see by (i) that the velocity of the particle is zero, that is, the particle at this instant is at rest. If t be taken greater than this the velocity becomes negative, showing that the particle is descending. Again, from equation (ii) we see that the space passed over during the time $\dfrac{u}{g}$ is $\dfrac{1}{2}\dfrac{u^2}{g}$; and since at this time the particle *begins* to descend, this is the greatest height to which it will rise; or, if h denote this greatest height, we have

$$h = \frac{1}{2}\frac{u^2}{g}.$$

Let w be the weight of the particle, m its mass; then, if we multiply each side of this last equation by w, that is mg, we get

$$wh = \frac{1}{2}mu^2,$$

which shows that the particle attains its greatest height and comes to rest when the work which it has done against its weight is numerically equal to the kinetic energy possessed by the particle at the beginning of the motion.

If we make t numerically greater than $\dfrac{u}{g}$, the value of s is less than $\dfrac{1}{2}\dfrac{u^2}{g}$, and when t is made equal to $\dfrac{2u}{g}$ we have $v = -u$ and $s = 0$, which shows that the particle will in the time $\dfrac{2u}{g}$ return to the point of projection, and on reaching that point its velocity will be downwards and numerically equal to u.

By comparing the expressions for v and s it will be seen that at any point in its descent the velocity of the particle is numerically the same as when passing through that point in its upward course.

157. **Ex. 2.** *A particle is allowed to slide from rest down a smooth inclined plane, whose inclination to the horizon is a, the only force acting upon it being its weight and the pressure of the plane. Find the motion.*

Let w be the weight of the particle, m its mass, then the resultant force upon the particle, that is, the force which tends to produce acceleration, is $w \sin a$, and acts directly down the plane. The particle will therefore move with a constant acceleration $\dfrac{w \sin a}{m}$, that is $g \sin a$, down the plane. Its velocity at the end of the time t will therefore be given by the equation

$$v = g \sin a \cdot t;$$

and the distance along the plane through which the particle will have fallen in t units of time will be given by

$$s = \frac{1}{2} g \sin a \cdot t^2.$$

Comparing these two equations we see that

$$v^2 = 2g \sin a \cdot s, \text{ or } \frac{1}{2} mv^2 = w \sin a \cdot s.$$

Now $s \sin a$ is the vertical height through which the particle has fallen, and we see that its velocity depends only on this vertical height and is independent of the inclination of the plane.

158. As a third example we may take the following.

A particle slides from rest down a rough inclined plane, whose inclination is a and coefficient of friction, when the particle is in motion, μ, determine the motion.

Let W be the weight of the particle, m its mass. If the friction be insufficient to support the particle, resolving the forces on the particle along the plane, we see that the resultant force acting down the plane is

$$W \sin a - \mu R.$$

Resolving perpendicularly to the plane we obtain
$$R = W \cos a.$$

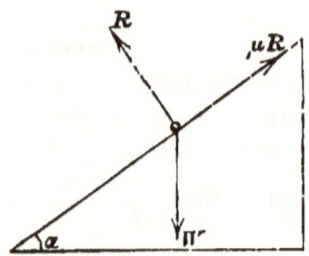

Hence the resultant force acting upon the particle is
$$W (\sin a - \mu \cos a).$$

It will therefore move down the plane with uniform acceleration
$$\frac{W}{m} (\sin a - \mu \cos a) \text{ or } g (\sin a - \mu \cos a).$$

Its velocity at the end of any time t will be
$$gt (\sin a - \mu \cos a),$$
and the space passed over during that time will be
$$\frac{1}{2} gt^2 (\sin a - \mu \cos a).$$

In this case the velocity of the particle, and therefore its kinetic energy, is less than that due to the vertical height through which it has fallen; the energy which is wanting is in this case transformed into heat by the friction.

159. *If a particle of mass* m *be moving with a velocity* v, *and be acted upon by a constant force* P *which brings it to rest in* t *units of time,* P *must be equal to* $\frac{mv}{t}$.

For since the force P is constant, the momentum destroyed by it in each unit of time is the same. But the momentum destroyed in t units of time is mv. Therefore the momentum destroyed in one unit of time is $\frac{mv}{t}$, which is therefore the measure of the force.

160. *A force* P *which is constant in magnitude and direction acts upon a particle of mass* m *while the particle moves from rest through a space* s. *Find the velocity generated by it.*

Since the force P is constant in magnitude and direction, the acceleration produced by it in a particle of mass m is constant, always in the direction of motion, and is given by the equation

$$P = mf.$$

The equation $v^2 = 2fs$ is therefore immediately applicable, and we have for the velocity required

$$v = \sqrt{\frac{2Ps}{m}}.$$

Similarly, if we have given the velocity generated from rest in a particle of mass m by a uniform force, while the particle moves over a distance s in the direction of the force, we can determine the magnitude of the latter; and if the particle, moving initially with a velocity v, be brought to rest by the force while moving over the space s, we can from the equation

$$\frac{1}{2} mv^2 = Ps$$

at once determine the magnitude of the force.

If however the particle does not start from rest, or is not brought to rest by the force, but only has its velocity increased or diminished while passing over the distance s, we must determine the magnitude of the force from the equation

$$\frac{1}{2} mv^2 - \frac{1}{2} mu^2 = Ps,$$

since the change produced in the kinetic energy of the particle is equivalent to the work done by the force.

161. As an example of the preceding article we will consider the following problem:—

A force equal to the weight of 1200 *tons acts upon an Armstrong shaft of* 700 *lbs. while it moves from rest through a distance of* 12 *ft. in the direction of the force, and the force then ceases. If the shaft without diminution of velocity strike a target and penetrate to a depth of* 12 *inches, then coming to rest, find the pressure exerted by the shot on the target, supposing it uniform during the penetration.*

A pound being taken as unit of mass the weight of a pound is equal to g absolute units of force. Hence, the mass of the shaft being denoted by 700, the force acting upon it will be $1200 \times 2240 \times g$ units of force.

The velocity of the shaft when the force ceases is determined from the equation of energy

$$\frac{1}{2} mv^2 = Ps,$$

and this becomes

$$\frac{1}{2} \cdot 700 \cdot v^2 = 1200 \cdot 2240 \cdot g \cdot 12;$$

$$\therefore v = \sqrt{\frac{1200 \cdot 2240 g \cdot 12}{350}} \text{ feet per second,}$$

$$= 96 \sqrt{10g} \text{ feet per second.}$$

Now this velocity is destroyed by a uniform force while the shaft moves through a distance of one foot in the direction opposite to that in which the force acts. Let P' be the measure of this force in absolute units. Then P' has to be determined from the equation

$$\frac{1}{2} mv^2 = P's.$$

And this becomes in this case

$$\frac{1}{2} \cdot 700 \cdot v^2 = P'$$

EXAMPLES. 123

Substituting for v^2 we get
$$P = 1200 \ . \ 2240 \ . \ 12 \ . \ g.$$
Hence the pressure exerted on the target is
$$1200 \ . \ 2240 \ . \ 12 \ . \ g$$
absolute units of force, that is 1200×12, or 14,400 tons' weight.

In this case the work done upon the target is equal to the work done originally upon the shot, and the mean pressure upon the target might have been at once determined from this consideration.

If it be required to find the time during which this pressure is exerted, we have simply to make use of the equation
$$v = \frac{P}{m} \ . \ t,$$
whence
$$t = \frac{700 \ . \ 96 \ . \ \sqrt{10 \ . \ g}}{1200 \ . \ 2240 \ . \ 12 \ . \ g} = \cdot 00116...,$$
or the time during which the pressure is exerted is about ·00116 of a second.

162. We will now work out, in full, a few examples illustrating the principles of this chapter; and it may be observed that much more of a subject may frequently be learned by a careful study of a few problems worked out in detail, than by any other method.

Ex. 1. *Two heavy particles* A *and* B, *whose masses are respectively* M *and* m, *of which* M *is the greater, are connected by a string of insensible mass passing over a smooth peg. Find the motion, and the space passed over by each in the first t seconds after the beginning of the motion.*

Let T be the tension of the string, which will be the same throughout, since the peg is smooth. Then the forces acting on the particle A are its weight Mg, acting downwards, and the tension T of the string acting upwards. The resultant force is therefore $Mg - T$ acting downwards, and the acceleration produced by this in the mass M will be
$$g - \frac{T}{M}.$$

124   EXAMPLES.

Again, the forces acting on the particle B, of mass m, are its weight mg acting downwards, and the tension. T,

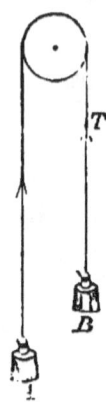

of the string acting upwards. The resultant force is therefore $T-mg$ acting upwards, and the upward acceleration produced by this force will be

$$\frac{T}{m} - g.$$

Now since the string is of invariable length and remains tight, the acceleration of B upwards must be equal to the acceleration of A downwards. Therefore

$$\frac{T}{m} - g = g - \frac{T}{M},$$

or $$T = \frac{2g \cdot Mm}{M+m};$$

and the acceleration of each particle is

$$g \frac{M-m}{M+m}.$$

Hence the velocity, v, of each particle at the end of time t is given by

$$v = \frac{M-m}{M+m} gt;$$

and the space s moved over by each particle during the time t is equal to

$$\frac{1}{2}\frac{M-m}{M+m}gt^2$$

The kinetic energy of the system is the sum of the kinetic energies of the two particles, that is

$$\frac{1}{2}Mv^2 + \frac{1}{2}mv^2$$

$$= \frac{1}{2}(M+m)\left(\frac{M-m}{M+m}gt\right)^2$$

$$= \frac{1}{2}\frac{(M-m)^2}{M+m}g^2t^2.$$

Also the work done by gravity upon the particle A is

$$Mgs = M\frac{1}{2}\frac{M-m}{M+m}\cdot g^2t^2,$$

and the work done *against* gravity in raising the particle B is

$$mgs, \text{ or } m.\frac{1}{2}\frac{M-m}{M+m}g^2t^2.$$

Hence the whole work done by gravity upon the system is

$$(M-m)\frac{1}{2}\frac{M-m}{M+m}g^2t^2,$$

that is

$$\frac{1}{2}\frac{(M-m)^2}{M+m}g^2t^2,$$

and is therefore equal to the kinetic energy of the system.

163. *In the preceding problem suppose the string to extend below the particle* B, *and when the system has been in motion for* t *seconds let a third particle* C, *of mass* m′, *initially at rest, be attached to the end of the string below the particle whose mass is* m, *the string remaining stretched throughout. Determine the motion and the impulse of the tensions of the portions of the string.*

126   MOTION OF BODIES CONNECTED BY A STRING.

The common velocity of the particles A and B at the

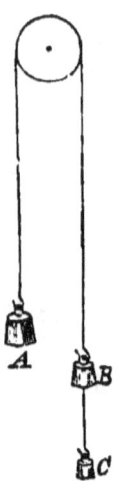

end of t seconds is, by the preceding investigation, $\dfrac{M-m}{M+m} gt$. Let this be denoted by v.

Now when C is attached, the string between B and C becomes suddenly tight, and C moves off with a jerk. Let u be the velocity with which C starts off, then the impulse of the tension acting upon C must be equal to $m'u$, which must therefore be the measure of the impulse of the tension of the string between B and C. Now immediately after C has been attached, the three particles A, B, and C must be moving with the same velocity, viz. u, since the string remains stretched (and we suppose the string inelastic). The velocity of A is therefore changed from v to u, u being, of course, less than v. The impulse of the jerk acting upon A must therefore be measured by $M(v-u)$, which is consequently the effect of the impulsive tension of the string between A and B; and this must be the same throughout. Hence an impulse represented by $M(v-u)$, due to the impulsive tension of the string above B, is exerted upon the particle B vertically upwards. But an impulse $m'u$, due to the impulsive tension of the string between B and C, is exerted upon B vertically downwards. The resultant impulse upon B is

MOTION OF BODIES CONNECTED BY A STRING. 127

therefore $m'u - M(v-u)$ vertically downwards. But the velocity of B is changed thereby from v to u, and B moves upwards throughout. Therefore

$$m'u - M(v-u) = m(v-u),$$
$$\text{or } (M+m+m')u = (M+m)v\,;$$
$$\therefore u = \frac{M+m}{M+m+m'}\,v$$
$$= \frac{M-m}{M+m+m'}\,gt.$$

Now the impulse of the tension of the string between A and B is $M(v-u)$, and substituting for v and u, we get for this impulse

$$\frac{Mm'(M-m)}{(M+m)(M+m+m')}\,gt.$$

Also the impulse of the tension of the string between B and C is $m'u$, that is

$$\frac{m'(M-m)}{M+m+m'}\,gt.$$

After the impulse the system is moving with velocity u, and it may be shown, as in the previous example, that the acceleration of A downwards and of B and C upwards is now

$$\frac{M-m-m'}{M+m+m'}\,g.$$

The velocity at the end of t' seconds after C was attached will therefore be

$$u + \frac{M-m-m'}{M+m+m'}\,gt',$$

and the space moved over during these t' seconds by each particle will be

$$ut' + \frac{1}{2}\frac{M-m-m'}{M+m+m'}\,gt'^2.$$

If $m+m'$ be greater than M the motion will be retarded after C is attached, and finally be in the opposite direction.

164. In the previous example the velocity, u, of the

system immediately after C was attached might have been determined from the consideration, that since no *impulsive* force external to the system of these particles acts upon it in the direction of the motion,* the momentum of the whole system immediately before and immediately after C is attached to the string must be the same, since the weights of the particles, which are finite forces, cannot generate a finite momentum in an indefinitely short time. But the momentum of C before being attached to the string was zero, since it was at rest. Hence we must have

$$(M + m) v = (M + m + m') u,$$

whence $$u = \frac{M+m}{M+m+m'} v,$$

as in the previous solution.

165. Before quitting this part of our subject we will consider one other example.

The two particles A *and* B, *whose masses are* M *and* m *respectively, being connected as in the previous examples, after the system has been in motion for* t *seconds, a third particle* C, *of mass* m′, *originally at rest, is attached to the string above* A ; *it is required to find the subsequent motion, the string being inelastic.*

As before, if v be the velocity of the system at the end of t seconds from the commencement of the motion, we shall have

$$v = \frac{M-m}{M+m} gt \quad \ldots\ldots\ldots\ldots\ldots\ldots(A).$$

Now when the particle C is attached above A, there will be an impulsive tension of the string between A and C, and C will move off with an initial velocity, which we will denote by u. Also, since the string between A and C is inelastic, it will remain stretched, and A and C will proceed with a common velocity, viz. u. The impulse of the jerk which must act upon C to make it start

---

* The pressure of the smooth peg is always at right angles to the motion of the part of the string on which it acts and therefore cannot change the velocity.

off with the velocity u is $m'u$, which is therefore the effect of the impulsive tension of the string between A and C. Hence the particle A receives an impulse upwards, de-

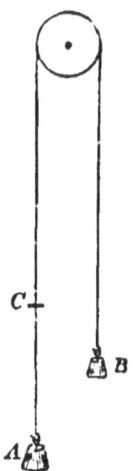

noted by $m'u$. But the velocity of A is changed by this impulse from v to u. Therefore

$$m'u = M(v-u),$$

or
$$u = \frac{Mv}{M+m'}$$

$$= \frac{M(M-m)}{(M+m')(M+m)} gt \ldots\ldots\ldots(B).$$

Now since the velocity of A is diminished, and there is no impulsive force acting upon B to diminish its velocity, immediately after the impulse B will be moving faster than C and A. Hence the string between C and B will become slack, and B will be moving as a particle acted upon by its own weight only, and projected vertically upwards with an initial velocity v. The velocity of B at the end of t' seconds after C is attached will therefore be $v - gt'$, and the height through which C will have risen in that time will be denoted by

$$vt' - \frac{1}{2} gt'^2.$$

While the string above C is slack, A and C will be falling freely, having started with an initial velocity u. Hence the common velocity of A and C at the end of the time t' will be
$$u + gt';$$
and the space through which they will have fallen during this time will be
$$ut' + \frac{1}{2}gt'^2.$$

Now the string will become tight when C has descended through a space equal to that through which B has ascended in the same time. Hence, if the string become tight at the end of t' seconds after C was attached, we must have
$$ut' + \frac{1}{2}gt'^2 = vt' - \frac{1}{2}gt'^2,$$
or
$$gt' = v - u,$$
whence
$$t' = \frac{v-u}{g}$$

and substituting for v and u their values from (A) and (B), we get
$$t' = \frac{M-m}{M+m}t - \frac{M(M-m)}{(M+m')(M+m)}t$$
$$= \frac{m'(M-m)}{(M+m')(M+m)}t \quad \ldots\ldots\ldots\ldots\ldots\ldots(C),$$

and the space through which the system will have moved since the attachment of C will be
$$\frac{M-m}{M+m}gt \cdot \frac{m'(M-m)}{(M+m')(M+m)}t - \frac{1}{2}g\left\{\frac{m'(M-m)}{(M+m')(M+m)}\right\}^2 t^2$$
$$= \left\{1 - \frac{1}{2}\frac{m'}{M+m'}\right\}\frac{m'(M-m)^2}{(M+m')(M+m)^2}gt^2.$$

MOTION OF BODIES CONNECTED BY A STRING. 131

Also the velocity of C and A when the string becomes tight will be

$$u + gt' = \frac{M(M-m)}{(M+m')(M+m)}gt + \frac{m'(M-m)}{(M+m')(M+m)}gt$$

$$= \frac{M-m}{M+m}gt,$$

the same as the velocity of the system before C was attached.

Also the upward velocity of B when the string becomes tight is

$$v - gt' = \frac{M-m}{M+m}gt - \frac{m'}{M+m'} \cdot \frac{M-m}{M+m}gt, \; \{\text{by (A) and (C)}\},$$

$$= \frac{M}{M+m'} \cdot \frac{M-m}{M+m}gt.$$

Now when the string is tight all the parts of the system must be moving with a common velocity. Hence when the string becomes tight there will be an impulsive tension. To determine the common velocity immediately after the tightening of the string, we may adopt the method of the last article. The momentum of the whole system, reckoned as though all the bodies were moving in the same direction, since the smooth peg simply serves to change the direction of the string, must be the same before and after the tightening of the string. Hence, if v' be the common velocity immediately after the string becomes tight, we must have

$$(M+m+m')v' = (M+m')\frac{M-m}{M+m}gt + \frac{Mm}{M+m'} \cdot \frac{M-m}{M+m}gt$$

$$= \frac{(M+m')^2 + Mm}{M+m'} \cdot \frac{M-m}{M+m}gt.$$

Therefore $\quad v' = \dfrac{(M+m')^2 + Mm}{(M+m')(M+m+m')} \cdot \dfrac{M-m}{M+m}gt.$

The effect of the impulsive tension of the string may be determined from the consideration that by it the velocity of the particle B, whose mass is m, is changed from

$$\frac{M}{M+m'} \cdot \frac{M-m}{M+m} gt \text{ to } v'.$$

After the string has become tight, since it is inelastic, it will remain tight, and the system starting with the velocity v will move with a constant acceleration equal to

$$\frac{M+m'-m}{M+m'+m} \cdot g ;$$

whence the velocity at any subsequent time, and the space passed over during any time, can be immediately found.

166. We will leave to the reader to investigate the motion of the system when, after the two weights A and B have been in motion for t seconds, the weight C, originally at rest, is suddenly attached to the string between B and the peg, and at a considerable distance above B. The reason for introducing the last condition will be readily seen. The whole problem is very similar to that last investigated, and will well repay the student for the time he will require to examine it.

167. If a point be moving with a known acceleration in the direction of its motion, and its initial velocity be given, the time in which it will describe any portion of its path can be immediately found from the equation

$$s = ut + \frac{1}{2} ft^2,$$

which is a quadratic equation for finding t.

If u and f be of the same sign, one of the roots of this equation is negative. The positive root is of course the one required. The negative root gives the time before the earliest time considered in the question, at which, if the point had been at a distance s in the positive direction from the point from which we have supposed it to start, and moving with a proper velocity in a direction opposite to its acceleration, it would have passed through the point from which we have supposed it to start, and have returned to it with a velocity u, at the instant at which we have supposed it to start.

If u and f be of opposite signs both the values of t are frequently admissible, as for example, in the case of a particle projected vertically upwards, when the smaller value of t gives the time in which it will reach a point vertically above the point of projection and at a distance S from it during its ascent, and the larger value of t gives the time at which it will reach the same point in its descent.

168. As an example of the preceding article we will consider the following problem.

A heavy particle falls from rest at the highest point of a vertical circle of diameter a down a smooth chord of the circle. Find the time of descent.

Let m be the mass of the particle; then mg will repre-

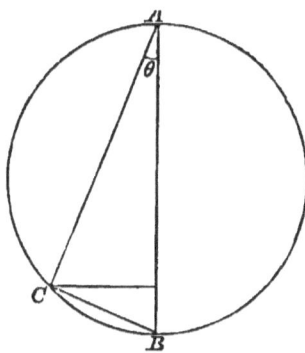

sent its weight. Let θ be the inclination of the chord to the vertical, s its length. Then

$$s = a \cos \theta.$$

Now the resultant force on the particle acts down the chord and is equal to $mg \cos \theta$. The acceleration produced by this force in the particle whose mass is m is $g \cos \theta$. Hence the particle moves down the chord with uniform acceleration $g \cos \theta$, and if t be the time of descent, we shall have

$$s = \frac{1}{2} g \cos \theta \cdot t^2,$$

or $\qquad a \cos \theta = \frac{1}{2} g \cos \theta \cdot t^2.$

Therefore

$$t = \sqrt{\frac{2a}{g}},$$

and is independent of the inclination of the chord.

Hence the time of descent down all chords of a vertical circle from the highest point is the same, and equal to the time of descent down the diameter.

Similarly the time of descent from any point of a vertical circle along the chord to the lowest point is the same as the time down the vertical diameter.

Since all the sections of a sphere by vertical planes through its highest point are equal circles, it follows that the times of descent of a particle down all chords from the highest point of a sphere are the same.

169. If it be required to find the straight line of quickest descent to any plane curve AB from a point P in its own

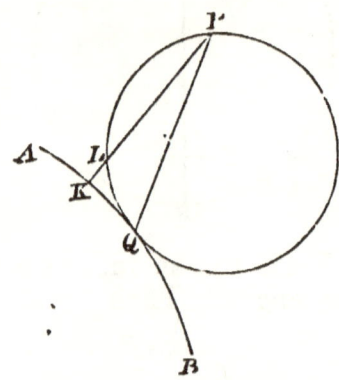

plane, we have only to describe a vertical circle having P for its highest point and touching the curve. Then the chord PQ drawn from P to the point of contact of the curve and circle is the line of quickest descent required. For if PQ be not the line of quickest descent, let some other straight line as PK be that of quickest descent. Let PK cut the circle in L. Then the time down PK is greater than that down PL. But the time down PL is equal to the time down PQ; therefore the time down PK is greater than that down PQ, which is contrary to the hypothesis that PK is the straight line of quickest descent from P to the curve.

If it be required to find the straight line of quickest descent from any point P to a given surface, it is only necessary to describe a sphere having P for its highest point and touching the surface. The chord drawn from P to the point of contact will then be the straight line of quickest descent from P to the surface.

170. If a particle be in motion under the action of a force always in the direction of its motion, since the work done by a force is measured by the product of the force and the distance moved over by its point of application in the direction of the force, it follows that the work done on the particle during any time is measured by the product of the force and the space passed over by the particle during that time. Hence the measure of the rate at which the force does work on the particle at any instant is the product of the force and the velocity of the particle at that instant, and is therefore proportional to the velocity. Hence if a particle move from rest with a constant acceleration in the direction of its motion, the rate at which work is done upon it is proportional to the time during which it has been moving.

We have said that the work done during any time τ by a uniform force acting upon a particle in the direction of its motion is measured by the product of the force and the space passed over by the particle during the time τ. Now if v' be the mean velocity of the particle during the time τ, v its velocity at the beginning, and $v + u$ its velocity at the end of the time, we have $v' = v + \frac{1}{2} u$, and the space passed over during the time τ will be represented by $v'\tau$. Hence the work done by the force during the time τ may be represented by the product of the force, the time τ, and the mean velocity v' of the particle. But the product of the force into the time is the measure of the momentum generated during that time. Hence the work done by the force during any time is measured by the product of the mean velocity of the particle and the momentum generated during that time.

Now the kinetic energy of a moving particle has been defined as one-half the product of its velocity and its momentum. It is therefore represented by $\frac{1}{2}mv \cdot v$, or $\frac{1}{2}mv^2$, if v be the velocity and m the mass of the particle. Now suppose that, during the time τ, v is changed to $v+u$. (By taking τ small enough u can be made as small as we please.) Then the kinetic energy becomes $\frac{1}{2}m(v+u)^2$. The increment of the kinetic energy is therefore $muv + \frac{1}{2}mu^2$. Hence we have for the increment of the kinetic energy the expression $mu\left(v + \frac{1}{2}u\right)$. Now mu is the momentum generated during the time τ and $v + \frac{1}{2}u$ is the mean velocity of the particle during that time, since the velocity changes uniformly. Hence the increment of the kinetic energy of the particle during any time is measured by the product of the momentum generated during that time and the mean velocity of the particle during the interval. The increment of the kinetic energy of the particle is therefore equivalent to the work done upon it by the force producing acceleration. From this investigation we see a reason for the definition we have given of kinetic energy.

171. If the force acting on the particle in the direction of its motion be not uniform, then if we make τ very small we may consider the force uniform during the interval τ, and the above investigation will hold. Hence in this case the increment of the kinetic energy produced in any very small time is equivalent to the work done during that time, and this being always true, it follows that the increment during any finite time, being the sum of the increments during the intervals into which that time may be divided, is equivalent to the work done on the particle during that time.

Hence if a particle move from *rest* under the action of

a force in the direction of its motion, whether its magnitude be constant or variable, the whole kinetic energy of the particle at any time will be equivalent to the whole amount of work done upon it by the force.

If the force be not in the direction of motion it may be resolved into two, one in that direction and one perpendicular to it. Now the latter component does no work, since its point of application moves always in a direction perpendicular to that of the force, and it may be shown that it does not increase the velocity of the particle, but simply tends to change the direction of its motion, and therefore does not alter its kinetic energy. Hence the former component is the only one which we need consider, and it follows that the whole change of the kinetic energy of the particle during any time is equivalent to the work done upon it during that time.

This result is a case of the Principle of the Conservation of Energy.

172. From the preceding articles it follows that if the resultant force acting on a particle at every point of its path be known, as well as the velocity at any given point, the velocity at any other point can be found if we can find the work done by the forces in passing from the one point to the other. Some examples illustrating this result will be found in Chapter V.

Since the rate at which work is done upon a particle by a force acting in the direction of its motion is measured by the product of the force and the velocity of the particle, it follows that if this rate be uniform, the force must vary inversely as the velocity of the particle. Hence if a particle start from rest, it is impossible for an agent to do work upon it at a finite rate at the commencement of the motion, since in order to do this, it would be necessary to exert an infinite force.

For example, it is impossible for an engine in starting a train to work up to its full horse-power.

173. We may illustrate this subject by the following examples.

EXAMPLES ON HORSE-POWER.

Ex. 1. *A train whose mass is* 100 *tons (including the engine) is drawn by an engine of* 150 *horse-power. The resistance to motion on a level line due to friction being equivalent to a force of* 14 *lbs.' weight for every ton in motion, and the resistance of the air being neglected, find the maximum speed which the engine is capable of sustaining on a level line.*

The resistance to the motion of the train due to friction, etc., is equal to the weight of 1400 lbs. Hence if v be the velocity of the train in feet per second, the rate at which its engine works is $1400v$ foot-pounds per second. But it is capable of doing 150×550 foot-pounds per second. Hence we must have, if v be the velocity required,

$$1400v = 150 \times 550,$$

$$v = \frac{15 \times 55}{14};$$

or the velocity of the train is $\dfrac{15 \times 55}{14}$ feet per second, that is $40\frac{5}{28}$ miles per hour.

174. Ex. 2. *If the train described in the preceding example be moving at a particular instant with a velocity of* 15 *miles per hour, and the steam turned off, how far will it run before coming to rest?*

The kinetic energy of the train is

$$\frac{100 \times 2240 \times 22^2}{2g}$$

foot-pounds, and the resistance to the motion is equal to the weight of 1400 pounds. The train will come to rest when the work done against the resistance is equal to the original kinetic energy. Hence if s feet be the distance which the train runs

$$1400\,s = \frac{100 \times 2240 \times 22^2}{2g}$$

$$\therefore s = \frac{100 \times 2240 \times 22^2}{2g \times 1400}$$

$$= 1210,$$

and the train will come to rest after running 1210 feet.

EXAMPLES ON HORSE-POWER. 139

175. Ex. 3. *Find the horse-power of an engine required to drag a train of* 100 *tons up an incline of* 1 *in* 50 *with a velocity of* 30 *miles per hour, the friction being equal to the weight of* 1400 *lbs.*

The line is inclined to the horizon at an angle whose sine is $\frac{1}{50}$. Therefore the resolved part of the weight of the train down the incline is 2 tons' weight. Hence the whole force tending to stop the motion of the train up the plane is equal to the weight of 2 tons and 1400 lbs., that is of 5880 lbs. Now a velocity of 30 miles per hour is 44 feet per second. Hence the engine must do 5880×44 foot-pounds of work per second, and the horse-power at which it works must therefore be $\dfrac{5880 \times 44}{550} = 470\frac{2}{5}$. The engine must therefore be of not less than 470^2 horse-power.

Since the pressure of the train upon the metals and bearings of the wheels when on an incline whose inclination is a, is less than the corresponding pressures on a level line in the ratio of $\cos a$ to 1, the friction will be less in the same ratio. Hence if the resistance to the motion of the train, due to friction when on an incline of 1 in 50 be 1400 lbs. weight, the corresponding resistance when on a level line will be

$$\frac{1400 \times 50}{\sqrt{50^2 - 1}} \text{ or } \frac{70000}{\sqrt{2499}}$$

pounds' weight.

176. Referring to the example in Art. 156, we see that if a particle be projected vertically upwards with velocity u it will rise to a height $\dfrac{u^2}{2g}$, and then come to rest. If the mass of the particle be denoted by m its weight will be represented by mg, and in falling from the height $\dfrac{u^2}{2g}$ it might be made to lift a weight, less than itself, but differing by as small a quantity as we please, through the same

height; that is, it may be made to do $mg \times \dfrac{u^2}{2g}$, or $\dfrac{mu^2}{2}$, units of work. Now the kinetic energy of the particle when projected is represented by $\dfrac{mu^2}{2}$, and we see that this amount of kinetic energy can be converted into the *capacity* or *power* of doing the same number of units of work due to the separation of the body from the earth, *i.e.* into the same number of units of potential energy.

If the particle be allowed to fall freely from the height $\dfrac{u^2}{2g}$, its kinetic energy on reaching the point of projection will be denoted by $\dfrac{mu^2}{2}$. Hence *kinetic energy* and *potential energy* are mutually convertible.

177. In estimating the work done by a force when its point of application is moved in a direction not coinciding with that of the force we may either suppose the force resolved into two components respectively parallel and perpendicular to the displacement, and since the former component is the only one by, or against, which work is done, we have only to multiply it by the displacement. Or we may resolve the displacement into two parts respectively parallel and perpendicular to the direction of the force and multiply the measure of the whole force by the component of the displacement in its own direction. Since forces and displacements are resolved according to the same law these two methods lead to the same expression for the work done; viz. $P \cdot d \cdot \cos a$, where P represents the force, d the displacement of its point of application, and a the angle between the direction of the force and that of the displacement.

178. In commenting on the Third Law of Motion we have given Newton's statement of the Principle of the Conservation of Energy. We have seen that the work done against conservative forces has its equivalent in potential energy generated in the system, while the work done in producing acceleration has its equivalent in the kinetic energy of the system. The destination of the

CONSERVATION OF ENERGY. 141

work done against friction and such like non-conservative forces was unknown to Newton and the energy was supposed to be lost. It is now known to be converted into heat or some other form of energy which does not come under our cognizance when treating of Mechanics. The Principle of the Conservation of Energy has been stated by Clerk Maxwell as follows :—

" *The total energy of any body or system of bodies is a quantity which can neither be increased nor diminished by any mutual action of these bodies, though it may be transformed into any of the forms of which energy is susceptible.*"

There is probably no law of nature which stands upon a firmer basis than this Principle of the Conservation of Energy, for there is no principle towards the overthrow of which so much ingenuity has been fruitlessly directed. The search for " the Perpetual Motion " has been carried on continuously for many generations, and the principle of the Conservation of Energy merely states that whatever agents may be called into action, " perpetual motion " or the *creation* of energy is impossible. (It should be noticed that *the* perpetual motion is quite distinct from the motion contemplated in Newton's first law, for it means perpetual motion *against resistance*, and therefore the continuous production of work.) We are therefore justified in assuming the truth of this principle in all cases as freely as we assume the first law of motion, and it will frequently be employed in this way in the following pages.

179. We have shown how to express the energy of a moving particle, whether given in foot-pounds or absolute units, in terms of its mass and velocity. If, then, in consequence of the constraints we can express the velocity of every particle of a system in terms of that of one of them or of any other single variable, and if we know all the external forces acting on a system, so that we can determine the amount of work done upon or by the system during any displacement, we can find at once the velocity of every particle of the system in any given configuration (compatible with the constraints to which

it is subjected) if we know the velocity in some standard configuration. The principle of the conservation of energy thus enables us to express the velocity of the parts of a system in terms of the displacement it has undergone without any reference to the time occupied in the displacement; but the time can be found subsequently, if required, from the knowledge of the change of motion and of the forces.

180. For example: *A body is projected in any direction with a velocity of* 100 *feet per second; find its velocity when it has reached a point* 20 *feet higher than the point of projection.*

The system with which we have here to deal consists of the earth and the projected body, but in consequence of the earth's mass we may neglect the work done upon the earth by the attraction of the body, and, as explained above, we shall speak of the kinetic energy of the body as equivalent to that of the whole system. Let m denote the mass of the body in pounds. Its energy is at first $\dfrac{m\,100^2}{2g}$ foot-pounds. The work done against gravity in rising 20 feet in $20m$ foot-pounds, and if v denote the velocity of the body in feet per second at this height, its kinetic energy is then $\dfrac{mv^2}{2g}$ foot-pounds.

Hence we must have
$$\frac{mv^2}{2g} + 20m = \frac{m\,100^2}{2g},$$
or
$$v^2 = 100^2 - 40g.$$
If $g = 32$ we have
$$v = 93\cdot38\ldots$$

181. *A particle moves under the action of a force which depends only on its position. The work done on the particle in passing from any point to any other point will be independent of the path pursued.*

Let the constraints consist of smooth surfaces or inextensible strings, since the direction of motion in these cases will be always at right angles to the constraining force, and therefore no work will be done by or against them.

Consider two paths A and B between the points P and Q. Then if the work done upon the particle in travelling from P to Q along the path A be not equal to that done in travelling along B, let that done in the path A be the greater and equal to W, while that done in the path B is denoted by W_1. Suppose the particle to travel from P to Q along A and return viâ B. Then in going from P to Q it will have W units of work done upon it by the external force and will, therefore, on reaching Q possess W units of kinetic energy. In returning from Q to P it will have to do W_1 units of work *against* the external force, since this force is independent of the direction of motion of the particle, depending only on its position. Hence since $W_1 < W$ when the particle reaches P it will have gained $W - W_1$ units of kinetic energy, and it will gain this additional quantity of energy every time it goes round the circuit, and we may abstract $W - W_1$ units of energy from the particle at the end of every revolution, and employ it to drive other machinery, and the motion will still be kept up. In fact we have "the perpetual motion" which is contrary to the principle of the conservation of energy. Hence $W - W_1 = 0$, or the work done in passing from P to Q is independent of the path.

This proof of course does not apply if energy is supplied to the system from an external source when the moving body describes a complete circuit, returning to its original position. For example, if a magnetic pole be capable of moving round a wire conveying an electric current, or if such a wire rotate around a magnetic pole, work is done during each revolution at the expense of the energy of the battery, and in such a case, in order to make the proof above given applicable we must introduce the condition that the path of the pole relative to the wire must not embrace the wire, in which case no work is done by the battery.

182. As a particular case of the preceding article, sup-

pose a particle to be moving under the action of its own weight only, and suppose Q to be vertically below P, the path A to be a vertical straight line, and the path B to be of any length and form whatever. The work done on the particle in sliding down B is the same as in falling down A, and therefore depends only on the *vertical* distance through which it descends.

Again, suppose Q not to be vertically below P. Let the path A consist of a horizontal and a vertical straight line, and let B as before be of any form. Then the work done in passing over the horizontal part of A is zero. Hence the work done in passing over B is the same as in traversing the vertical portion of A, and depends only on the vertical height through which the particle descends.

183. The following examples on units may be of use.

EXAMPLE 1. *A particle moves from rest through 300 feet with an acceleration denoted by 3 in 5 seconds, and acquires a velocity denoted by 30; what are the units of space and time?*

Let L feet be the unit of length and T seconds the unit of time. The unit of velocity will then be $\dfrac{L}{T}$ feet per second, and the unit of acceleration $\dfrac{L}{T^2}$ foot-second units.

The acceleration denoted by 3 will therefore be $3\dfrac{L}{T^2}$ foot-second units, and the velocity denoted by 30 will be $30\dfrac{L}{T}$ units.

Hence by the question

$$\frac{1}{2} \cdot 3 \frac{L}{T^2} \cdot 5^2 = 300,$$

and
$$3 \frac{L}{T^2} \cdot 5 = 30 \frac{L}{T};$$

$$\therefore \frac{1}{T} = 2 \text{ or } T = \frac{1}{2},$$

and
$$L = 2.$$

184. EXAMPLE 2. A *train of* 300 *tons is travelling at* 60 *miles an hour; find the measure of the force which will stop it in* 500 *yards, a hundred yards being the unit of length, a hundredweight the unit of mass, and ten seconds the unit of time.*

The mass of the train will be represented in the given system of units by 6000; its velocity by $\frac{880}{300}$ or $\frac{44}{15}$ (remembering that 60 miles an hour is 88 feet per second). Hence its kinetic energy will be represented by

$$\frac{1}{2} \cdot 6000 \frac{44^2}{15^2}.$$

This energy is transformed while the train runs 500 yards, that is, 5 units of length. Therefore the measure of the force is

$$\frac{1}{5} \cdot \frac{1}{2} \cdot 6000 \frac{44^2}{15^2} = 5162\frac{2}{3}.$$

EXAMINATION ON CHAPTER II.

1. When is the velocity of a moving particle said to be uniform? If the velocity of a particle be uniform and denoted by 6, when a mile is the unit of length and an hour the unit of time, find the space which it will pass over in 10 seconds.

2. Find the distance through which a particle will fall from rest under the action of gravity in 4 seconds.

If the depth of the surface of the water in a well be 256 feet and the velocity of sound 1120 feet per second, find the time which must elapse after dropping a stone before hearing it strike the water.

3. What is the numerical measure of the kinetic energy of a pound which has fallen freely from rest through a vertical height of 80 feet?

4. If in Attwood's machine the two weights P and Q be each one pound, and the weight R one ounce, and if

G. D.　　　　　　　　　　　　　　　　　　　　　L

the height through which R falls be 2 feet, find the subsequent velocity of P and Q.

5. If a particle which has fallen freely from rest move in one particular second through 176 feet, find how long it had been falling before the beginning of that second.

6. A particle slides down a smooth plane inclined at an angle of 30° to the horizon. Find its velocity after moving from rest through 10 feet, and the time occupied in moving over that distance.

7. Find the time occupied by a particle in sliding through 20 feet down a rough plane inclined 60° to the horizon, the coefficient of friction being one half: and its velocity at the end of the time.

8. Show that the difference of the kinetic energy of two particles after sliding through the same distance down two planes equally inclined to the horizon, but one rough and the other smooth, is numerically equal to the work done against friction by the former particle.

9. A particle is projected vertically upwards with a velocity of 100 feet per second; find the time occupied by it in its ascent in describing that portion of its path which lies between the heights of 60 ft. and 120 ft. above the point of projection.

10. A point P is at a distance of 12 feet from a plane inclined 30° to the horizon, and is above the plane. Find the time of quickest descent in a straight line from P to the plane.

11. Find the measure of the least impulse with which a particle of mass m can be projected to a vertical height of 400 feet.

12. If the unit of impulse be that required to project one pound vertically up one foot, find the measure of the impulse with which 4 lbs. can be projected vertically up 4 feet.

EXAMPLES ON CHAPTER II.

1. A particle descends an inclined plane; if the upper portion be smooth and the lower rough, the coefficient of friction being μ, and if the smooth length be to the rough length as $p:q$, show that the particle will just come to rest at the foot of the plane if $\mu = \dfrac{p+q}{q}\tan a$, where a is the inclination of the plane to the horizon.

2. A point moving with uniform acceleration f in the direction of motion describes σ feet in τ seconds, and at the end of t seconds is moving with velocity v. Find the units of time and space.

3. Find the line of quickest descent from one given circle to another, the circles being in the same vertical plane.

4. The resistance to the motion of a train due to friction, etc., being equal to the weight of 12 lbs. per ton, if the train moving at the rate of 40 miles an hour come to the foot of an incline of 1 in 200, the steam being turned off, find how far the train will go before it comes to rest.

5. If, in the preceding example, the train come to the top of the same incline, find how far it will descend the incline.

6. The resistance to the motion of a train will just allow it to run with uniform velocity down an incline of 1 in 150. If the train come with a velocity of 20 miles an hour to the top of an incline of 1 in 100, and run down this incline for one mile with the steam turned off, find how far it will be able to run along a level line from the base of the incline without turning on the steam.

7. If the resistance to the motion of a train on a level line be equal to the weight of 12 lbs. per ton of its mass, and if a train of 120 tons running on a horizontal line

at the rate of 40 miles an hour be brought to rest in a quarter of a mile by the application of a break to each wheel of a break van of 10 tons, the break entirely preventing the rotation of the wheels, find the coefficient of friction between the wheels and rails.

How far would the train have run if the break had not been put on, and how far if the breaks had been applied to each of two ten ton break vans?

N.B. In working this example the "rolling friction" amounting to 120 lbs.' weight for each break van is not supposed to act when the breaks are on.

8. A shot of 300 lbs. is fired from a thirteen ton gun with a velocity of 1600 feet per second, the mass of the powder being 50 lbs.; supposing the products of combustion of the powder to leave the gun with the same velocity as the shot, find the velocity with which the gun will recoil, and the height to which it will run up a smooth plane inclined 15° to the horizon.

9. Compare the amount of work done on the gun in the preceding example by the explosion of the powder, with that done on the shot.

10. A bullet of 250 grains is fired from a rifle whose mass is 10 lbs. with a velocity of 1200 feet per second by the explosion of $2\frac{1}{2}$ drams of powder. Supposing the mean velocity with which the products of combustion of the powder leave the rifle to be four-fifths that of the shot, and supposing the rifle to be brought to rest by a uniform force while it kicks through 3 inches, find the force.

What would be the effect of placing the shoulder against a tree while firing a rifle?

11. If S be the focus of a parabola whose axis is horizontal and plane vertical, and if SP be the straight line of quickest descent from S to the curve, show that SP is inclined at an angle of 60° to the axis.

12. A particle is projected vertically upwards with a velocity represented by 160; find the time of its reaching a height of 256 ft., and its velocity at that height.

EXAMPLES. 149

13. A ball is projected vertically upwards with a given velocity, and when it has attained half its greatest altitude another ball is projected vertically upwards with the same velocity from the same point. Determine when and where they will meet.

14. A body describes 75 feet from rest and acquires a velocity denoted by 20, with a uniform acceleration denoted by 8, in 5 seconds. What are the units of space and time?

15. If the number of units of space described by a body in the last second of its fall be to the number of units in its final velocity as $8 : 9$, during how many seconds has the body been falling?

16. Find the statical measure of a force which in half a mile would stop a railway train of 120 tons, moving at the rate of 25 miles an hour.

17. AP, AQ are two inclined planes of which AP is rough, the coefficient of friction being equal to tan PAQ, and AQ is smooth, AP lying above AQ: show that if bodies descend from rest at P and Q they will arrive at A, (1) in the same time if PQ be perpendicular to AQ, (2) with the same velocity if PQ be perpendicular to AP.

18. A particle falls freely from A, and a second particle is allowed to fall from B, vertically below A, at the instant when the first particle is half-way between A and B. Find when and where they will be together, and show that their velocities will then be as $3 : 1$.

19. A point moving with uniform acceleration describes 20 feet in the half second which elapses after the first second of its motion; compare its acceleration with that of a heavy falling particle; and find its numerical measure, taking a minute as the unit of time and a mile as that of length.

20. A particle has been falling for 40 seconds; find the force which will stop it in 10 seconds; find also the force which will stop it in 10 feet.

21. A particle falling from rest through a feet, with

acceleration f, in t seconds, acquires a velocity b; what are the units of time and length?

22. A body whose weight is W descends vertically, and draws an equal body up a smooth plane inclined at an angle of 30° to the horizon, the two bodies being connected by an inextensible string passing over the edge of the plane; find the velocity of each body at a given time, and the tension of the string.

If the inclined plane be the upper surface of a wedge resting on a smooth horizontal table, find the force necessary to prevent the wedge from sliding along the table.

23. Two particles are connected by a string which hangs over a smooth pulley at the top of a smooth plane inclined 30° to the horizon, one of the particles hanging vertically and the other resting on the plane. If the acceleration of either be represented by $\frac{g}{2}$, find the ratio of their masses.

24. A particle of a mass m, acted on by a constant force, has a velocity with which, if it were free, it would describe 300 feet in one minute, and after 3 seconds it has a velocity of 2 feet per second. Find the force referred to a minute as the unit of time, a foot being the unit of length.

25. A series of particles slide down the smooth faces of a pyramid, starting simultaneously from the vertex: show that after any time t they are all on the surface of a certain sphere whose radius is $\frac{1}{4} g t^2$.

26. A particle is projected with a velocity u, and after describing a certain space has acquired an additional velocity v, but to acquire a second additional velocity v it must describe further a space double that which it described in the first instance. Show that $u = \frac{v}{2}$.

27. A particle of mass $4P$ is drawn up a smooth plane, inclined 30° to the horizon, by a second particle of mass

EXAMPLES. 151

$3P$, connected with the former by a string passing over the upper edge of the plane: find the tension of the string, and the time before the first particle arrives at the top of the plane.

28. During the 1st, 3rd, 5th,... seconds of a particle's motion it is subject to uniform accelerations f, $3f$, $5f$,... in the direction of motion, and during the 2nd, 4th, 6th,... seconds its acceleration is zero. Supposing the particle to start from rest, find the velocity acquired, and the space passed over, in $2n$ seconds.

29. A parabola is placed in a vertical plane with its axis inclined to the vertical. S is the focus, A the vertex, and Q the point on the curve which is vertically below S. If SP be the straight line of quickest descent from the focus to the curve, show that the angle ASP is equal to twice the angle PSQ.

30. Two given points are in the same vertical line: show that the locus of the points in a vertical plane through them, from which the times of descent down straight lines to the points are the same, is a rectangular hyperbola.

31. Two particles are let fall at different instants down two smooth inclined planes from the same point of the horizontal ridge in which they meet. Prove that each describes, relative to the other, a parabola.

If the direction of the relative motion make an angle θ with the plane which bisects the external angle between the two planes, then

$$2\theta = \alpha \sim \beta,$$

where α and β are the inclinations of the given inclined planes to the horizon.

32. A body moving down a smooth inclined plane is observed to fall through equal spaces, a, in consecutive intervals of time τ_1, τ_2; prove that the inclination of the plane to the horizon is

$$\sin^{-1}\left\{\frac{2a}{g\tau_1\tau_2} \cdot \frac{\tau_1 - \tau_2}{\tau_1 + \tau_2}\right\}.$$

33. Two weightless planes, inclined respectively at angles a and β to the horizon, are placed back to back on a rough horizontal table, and two weights P and Q, whose masses are m and m' respectively, are placed upon them, being connected by a string passing over the common vertex of the planes. If P predominate, and the motion take place in a vertical plane, find the pressure on the planes, and show that they will slide on the table if the coefficient of friction be less than

$$\frac{(m \sin a - m' \sin \beta)(m \cos a + m' \cos \beta)}{(m + m')^2 - (m \sin a - m' \sin \beta)^2}.$$

34. Two particles start simultaneously from the same point and move along two fixed straight lines, the one with uniform velocity, the other from rest with uniform acceleration. Prove that the line joining the particles at any time is a tangent to a fixed parabola.

35. Show that if two particles of mass m and m' respectively be suspended by a string over a smooth pulley, the acceleration of their common centre of gravity is

$$g \left(\frac{m - m'}{m + m'}\right)^2$$

36. Two particles are connected by a string: one of them rests on a smooth horizontal table and the other hangs over the edge. The system being allowed to move, determine the motion of the centre of gravity of the two.

37. Two particles are connected by a string which passes over a smooth pulley fixed at the top of two smooth inclined planes having a common height: supposing that one particle moves on each plane, and the whole motion is in one vertical plane, find the motion of their centre of gravity.

38. In a system of n pulleys, in which each hangs by a separate string, the power and weight are in equilibrium, the power consisting of a heavy suspended body. If the weight be doubled, show that its acceleration is $\dfrac{g}{2^n + 2}$, the pulleys being without weight.

39. Show that the straight line of quickest descent from one curve to another makes the same angle with the normal to either curve at the point where it meets the curve.

40. A number of equal weights are attached at different points to a string, and the string then hangs over a smooth pulley. Show that at any subsequent time the tensions of the successive portions of the string are, on each side, in Arithmetical progression.

41. A large number of equal particles are attached at equal intervals, a, to a string, and the whole is heaped up near the edge of a smooth table, the particle at one extremity of the string being just over the edge of the table. Show that, if v be the velocity of the system just before the $(n+1)^{th}$ particle is set in motion, v is given by the equation

$$v^2 = \frac{ga}{3} \cdot \frac{(n+1)(2n+1)}{n}.$$

Hence show that in the limit, when a is indefinitely small and na finite, the chain will descend with uniform acceleration $\frac{g}{3}$.

42. The down express approaching Abbot's Ripton, with all breaks on and steam shut off, reduced its speed from 60 to 20 miles per hour in 800 yards. How much additional space would have saved the collision (the other train being at rest)?

43. If the kinetic energy of a train of 100 tons moving at 45 miles an hour be represented by 11, while its momentum is represented by 5, and 40 horse-power is represented by 15, find the units of length, mass and time, and show that the acceleration produced by gravity will be represented by 2016, assuming its measure to be 32 in the foot-second system.

44. Two masses of 3 lbs. and 5 lbs. connected by a string which passes over a smooth pulley are in motion: if the acceleration of the system be the unit of accelera-

tion, and the velocity at the end of 10 seconds from the commencement of the motion be the unit of velocity, find the units of space and time.

45. An ellipse is placed with its minor axis vertical; prove that the normal chord of quickest descent from the curve to the major axis is that drawn from a point subtending a right angle at the foci, if there be such a point. What is the condition that there may be, and what is the normal chord of quickest descent if there is not?

46. A train, of mass 200 tons, is ascending an incline of 1 in 100 at a rate of 30 miles per hour, the resistance of the rails, etc., being equal to 1600 lbs.' weight. The steam being shut off and the break applied, the train is stopped in a quarter of a mile. Find the weight of the break-van, the coefficient of sliding friction of iron on iron being $\frac{1}{6}$.

47. Show that the locus of points reached in a given time by particles sliding from rest down equally rough straight lines, originating in a given point, is a spindle, or surface formed by the revolution of a segment of a circle about its chord.

48. A blacksmith, wielding a 14 lb. sledge, strikes an iron bar 25 times per minute, bringing the sledge to rest upon the iron after each blow. If the velocity of the sledge on striking the iron be 32 feet per second, compare the rate at which the smith works with a horse-power.

49. A steam pump raises 1000 gallons of water per minute, and delivers it, at a height of 300 feet through a circular orifice four inches in diameter. Neglecting the friction of the pipe, find the H.P. at which the engine must work, and determine how much of the power would be saved if the water were delivered through an orifice sixteen inches in diameter.

CHAPTER III.

ON PROJECTILES.

185. In this Chapter we propose to consider the motion of a heavy particle projected in a direction inclined to the vertical and acted upon by its own weight only, together with some problems immediately connected with this subject.

This is the first case which has come specially under our notice, in which the force acting upon the moving particle, and, therefore, its acceleration, are not in the straight line in which it is moving. The force is in this case always inclined to the direction of motion, and therefore tends continually to alter this direction, for the velocity generated in each short interval of time, being in the direction in which the force acts, is inclined to the direction in which the particle is moving at the beginning of the interval. The velocity at the end of the interval will be found by compounding these two velocities according to the parallelogram law, and its direction will be intermediate between those of the two components, and will therefore tend to coincide more nearly with the direction of the force. Hence the direction of motion of the particle changes continuously, tending always towards the direction of the force, and the particle will therefore describe a continuous curve with its concavity downwards. In the following articles we shall neglect the rotation of the earth and the friction of the air.

186. *A heavy particle is projected with velocity* u *in a direction making an angle* a *with the horizontal plane through the point of projection; to determine the subsequent motion and the point where the particle again reaches this horizontal plane.*

PATH OF A PROJECTILE A PARABOLA.

The whole motion will obviously be in a vertical plane passing through the line of projection, since there is no force tending to make the particle move out of this plane. Also since the resistance of the air is neglected, the only force acting upon the particle is its weight acting vertically downwards. Let m be the mass of the particle, w its weight, and let the plane of the paper represent the vertical plane through the line of projection. Let B be the point of projection; BT the direction of projection, BC horizontal, and let the curve $BPAC$ represent the

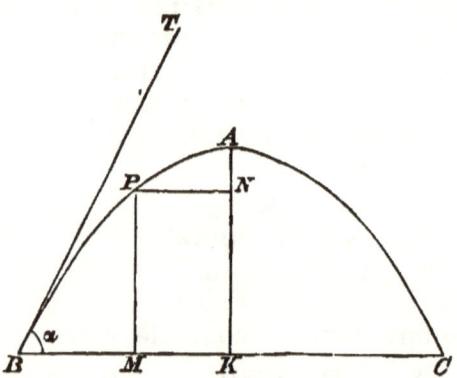

path of the particle. Then since BT is the direction in which the particle begins to move, BT must be the tangent to the path at B. Suppose the velocity of projection, u, to be resolved into two components, viz. $u \sin a$ vertically upwards and $u \cos a$ in the direction BC.

Now the only force acting on the particle is its weight, which acts vertically downwards, and since the change of motion produced by a force at each instant is in the direction of the force, the velocity generated in the particle at each instant of its flight is in the vertical direction, and its horizontal velocity remains unaltered throughout, and is therefore always equal to $u \cos a$.

Again, the acceleration produced by the weight is such as to generate g units of velocity in the vertical direction during each unit of time. Hence the vertical component of the velocity of the particle at the end of t units of time

PATH OF A PROJECTILE A PARABOLA. 157

after the projection is $u \sin a - gt$. Also the horizontal velocity of the particle can have no effect in altering its height above BC. Hence the vertical height through which the particle will have risen during the time t is, by equation (2) Art. (124), equal to

$$u \sin a \cdot t - \frac{1}{2}gt^2.$$

Also the distance through which the particle will have moved parallel to BC in the time t is $u \cos a \cdot t$.

Let P be the position of the particle at the end of any time t, PM vertical, then $PM = u \sin a \cdot t - \frac{1}{2}gt^2$, and $BM = u \cos a \cdot t$, and thus the position of P is determined.

If we make t equal to $\dfrac{u \sin a}{g}$, we see that the vertical component of the velocity of the particle is zero. The particle will therefore at the end of this time be moving horizontally. Let A be its position; then the tangent to the path at A is horizontal. Draw AK vertical. Then we have

$$AK = u \sin a \cdot \frac{u \sin a}{g} - \frac{1}{2}g\left(\frac{u \sin a}{g}\right)^2$$

$$= \frac{1}{2}\frac{u^2 \sin^2 a}{g}.$$

Also

$$BK = u \cos a \cdot \frac{u \sin a}{g}$$

$$= \frac{u^2 \cos a \sin a}{g} = \frac{1}{2}\frac{u^2 \sin 2a}{g}.$$

Draw PN parallel to BC. Then $PN = BK - BM$

$$= \frac{1}{2}\frac{u^2 \sin 2a}{g} - u \cos a \cdot t,$$

and $AN = AK - PM$

$$= \frac{1}{2}\frac{u^2 \sin^2 a}{g} - u \sin a \cdot t + \frac{1}{2}gt^2.$$

Now $PN^2 = \dfrac{u^4 \sin^2 2a}{4g^2} - \dfrac{u^3 \sin 2a \cos a}{g} t + u^2 \cos^2 a \cdot t^2$;

$\therefore PN^2 = \dfrac{2u^2 \cos^2 a}{g} \cdot AN.$

Also since this is independent of t it is true for all positions of P. Hence the path is a parabola of which A is the vertex, AK the axis, and whose latus rectum is equal to

$$\dfrac{2u^2 \cos^2 a}{g} \quad\quad\quad\quad\quad\quad\text{(A).}$$

Since when the particle is at A its motion is horizontal, after passing A it will begin to descend. Hence AK is the greatest height to which it will rise. Therefore if h be the greatest height to which the projectile will rise,

$$h = \dfrac{u^2 \sin^2 a}{2g} \quad\quad\quad\quad\quad\quad\text{(B).}$$

If P be taken on the side of AK remote from B, we must write for PN, $u \cos a \cdot t - \dfrac{u^2 \sin 2a}{2g}$; hence PN^2 is the same as before, and AN being of the same form as before, we have still

$$PN^2 = \dfrac{2u^2 \cos^2 a}{g} \cdot AN,$$

which shows that after passing A the particle continues to move in the same parabola as before. Hence KC is equal to BK, and

$$BC = 2BK$$
$$= \dfrac{u^2 \sin 2a}{g}.$$

Hence, if r denote the range on the horizontal plane through the point of projection, we have

$$r = \dfrac{u^2 \sin 2a}{g} \quad\quad\quad\quad\quad\quad\text{(C).}$$

PATH OF A PROJECTILE A PARABOLA. 159

The focus will be in AK at a depth below A equal to one-fourth of the latus rectum. Hence the height of the focus from the ground is

$$AK - \frac{u^2 \cos^2 a}{2g}$$

$$= \frac{u^2 (\sin^2 a - \cos^2 a)}{2g}$$

$$= -\frac{u^2 \cos 2a}{2g} \quad\ldots\ldots\ldots\ldots\ldots\ldots(D).$$

The directrix will be a line parallel to BC and at a height *above* A equal to one-fourth of the latus rectum. Hence the height of the directrix above the ground is

$$\frac{u^2 (\sin^2 a + \cos^2 a)}{2g}$$

$$= \frac{u^2}{2g} \ldots\ldots\ldots\ldots\ldots\ldots\ldots(E),$$

and it is therefore at such a height that if the particle were projected vertically with velocity u it would just reach the directrix. Hence if a number of particles be projected from B in the same vertical plane and with the same velocity but in different directions, the parabolas described by them will all have the same directrix.

From the principle of the conservation of energy combined with the fact that the velocity of projection is such that if a body were projected vertically with this velocity it would just reach the directrix, it follows that at any point of its path the velocity of a projectile is numerically equal to that which it would gain in falling vertically from rest at the directrix to the point at which it is. See Art. 189.

187. The proof that the path of a projectile is a parabola is generally given somewhat as follows.

Let the plane of the paper be the vertical plane of projection, BT the direction of projection making an angle a with the horizontal line BC. Let u be the velocity of projection. Suppose the velocity of the particle always resolved into two components, one in the direction BT, and the other vertically downwards. Now since the only

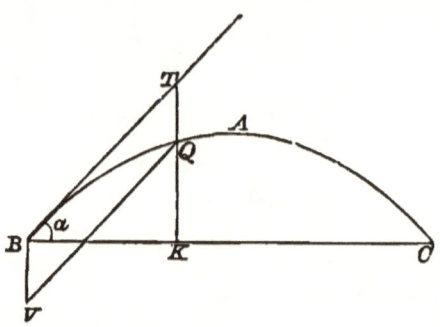

force acting on the particle is its weight, which acts vertically downwards, the velocity generated in each instant is always vertical, and due to a constant acceleration g. Hence during any particular instant it is only the vertical component of the particle's velocity which is affected, and this being true for each instant, it follows that throughout the motion the vertical component of the velocity is uniformly accelerated while the component in the direction BT remains the same, and is therefore always equal to u. The vertical component of the velocity which is zero at the commencement of the motion will be denoted by gt at the end of t seconds.

Now since the particle moves with a constant velocity u in the direction BT, and this velocity is independent of its vertical velocity, it will, in t seconds, on account of this component of its velocity alone have moved through a space ut. Take BT equal to ut and draw TK vertical. Then since the other component of the particle's velocity is vertical, the space passed over during *each* instant by the particle on account of this second component of its velocity is always vertical, and this being true for *each* instant during its flight, it follows that the space passed over by the particle during any time t on account of the second component of its velocity only is in the vertical direction. Hence at the end of the time T the particle must lie somewhere in the straight line TK. Let Q be its position. Now the distance TQ will be equal to the space passed over by the particle in the time t, on account of the vertical component of its velocity. But this component at the end of any time t is equal to gt; hence the space passed over from rest by the particle in t seconds moving with this component of its velocity only is $\frac{1}{2}gt^2$.

Therefore $TQ = \frac{1}{2}gt^2$. Draw BV vertical, and QV parallel to BT. Then $QV = BT' = ut$, and $BV = QT = \frac{1}{2}gt^2$,

$$\therefore QV^2 = u^2 t^2 = \frac{2u^2}{g} \cdot \frac{1}{2}gt^2$$

$$= \frac{2u^2}{g} BV,$$

and $\frac{2u^2}{g}$ is a constant quantity. But in the parabola $QV^2 = 4SP \cdot PV$ (Besant's *Conics*). Therefore, the curve is a parabola whose axis is vertical and vertex upwards, and the distance of B from the focus is $\frac{u^2}{2g}$, which is therefore also its distance from the directrix.

188. Assuming the path of a projectile to be a parabola, we can at once find the latus rectum and the position of the focus. The letters used to designate the equations are the same as in Art. 98.

Resolving the velocity always into two components, the first horizontal and the second vertical, these two components at the end of t seconds from the commencement of the motion are respectively $u \cos a$ and $u \sin a - gt$. The point is moving horizontally, and is therefore at the vertex A (see figure, Art. 186) of the parabola when its vertical velocity is zero, that is, when $t = \frac{u \sin a}{g}$. We have then

$$AK = u \sin a \cdot t - \frac{1}{2}gt^2$$

$$= \frac{u^2 \sin^2 a}{2g} \quad \dots\dots\dots\dots\dots(B),$$

G. D.

and
$$BK = u \cos a \cdot t$$
$$= \frac{u^2 \cos a \sin a}{g}.$$

But since the curve is a parabola, if l be the latus rectum we must have
$$BK^2 = l \cdot AK;$$
$$\therefore l = \frac{2u^2 \cos^2 a}{g} \quad \ldots\ldots\ldots\ldots\ldots(A).$$

The range is equal to twice BK, since the parabola is symmetrical about AK. Hence if r be the range,
$$r = \frac{2u^2 \cos a \sin a}{g}$$
$$= \frac{u^2 \sin 2a}{g} \quad \ldots\ldots\ldots\ldots\ldots(C).$$

In order that the range may be a maximum we must have $\sin 2a = 1$, and therefore $a = 45°$. Hence for a given velocity of projection an elevation of $45°$ gives the greatest horizontal range.

189. The directrix being at a height above A equal to one-fourth of the latus rectum, its height above BC is
$$\frac{u^2}{2g} \ldots\ldots\ldots\ldots\ldots(E).$$

Now the height of Q above BC at the end of the time t is $u \sin a \cdot t - \frac{1}{2}gt^2$. Hence the distance of Q below the directrix is
$$\frac{u^2}{2g} - u \sin at + \frac{1}{2}gt^2 \ldots\ldots\ldots\ldots(F).$$

Also the velocity of the particle is the resultant of its horizontal velocity, $u \cos a$, and its vertical velocity, $u \sin a - gt$, and these are at right angles. Hence if v denote the velocity of the projectile at the end of the time t,
$$v = \{u^2 \cos^2 a + (u \sin a - gt)^2\}^{\frac{1}{2}}$$
$$= \{u^2 - 2u \sin a \cdot gt + g^2t^2\}^{\frac{1}{2}} \ldots\ldots\ldots\ldots(G).$$

EXAMPLES OF PROJECTILES. 163

Now the distance, S, of Q below the directrix is given by

$$S = \frac{u^2}{2g} - u \sin a \cdot t + \frac{1}{2} gt^2,$$

and the velocity, V, of a particle falling freely through this vertical height is given by the equation.

$$V^2 = 2gS;$$
$$\therefore V^2 = u^2 - 2u \sin a \cdot gt + g^2 t^2.$$

Hence the velocity of the projectile at any time is that due to falling from the directrix to its position at the time. This is in accordance with the principle, that the increase of the kinetic energy of the body is equivalent to the work done by gravity upon it.

190. DEF. *The angle which the direction of projection makes with the horizontal plane through the point of projection is called the elevation of projection.*

We will now apply the methods and results of the preceding Articles to some examples.

A bullet is projected with a velocity of 1000 feet per second at an elevation of 15°. *Find its range on the horizontal plane through the point of projection, neglecting the resistance of the air.*

The range, r, is given by the equation

$$r \text{ feet} = \frac{u^2 \sin 2a}{g} \text{ feet}$$

$$= \frac{1000000 \cdot \frac{1}{2}}{g} \text{ feet}$$

$$= \frac{1000000}{64} \text{ feet,}$$

assuming g equal to 32.

The range is therefore 5208⅓ yards. Of course in the case of bodies moving with such great velocities as 1000 feet per second, the resistance of the air is very great

164   EXAMPLES OF PROJECTILES.

indeed, and the above result is practically worthless. The effect of the resistance of the air upon cannon shot is such, that the maximum range is obtained when the elevation of the line of fire is about 33°.

191. *A particle projected at an elevation a, and with velocity* u, *strikes a fixed vertical plane perpendicular to the vertical plane of projection, and at a distance* k *from the point of projection. Find the height at which it strikes.*

The horizontal velocity of the particle remains constant and equal to $u \cos a$. Hence if t be the time elapsing

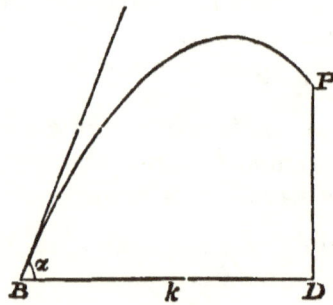

after the instant of projection and before the particle reaches the plane, we have

$$t = \frac{k}{u \cos a}.$$

The height to which the particle will rise in this time is given by the equation

$$s = u \sin a \cdot t - \frac{1}{2} g t^2,$$

or substituting for t,

$$PD = k \tan a - \frac{1}{2} g \frac{k^2}{u^2 \cos^2 a},$$

which determines the point where the particle strikes.

192. *Suppose it required to find the point where a projectile will strike a horizontal plane at a distance* h *below the point of projection.*

Let, as before, u be the velocity and a the elevation of projection. The time t in which the particle will rise to a height s above the point of projection is given by the equation

$$s = u \sin a \cdot t - \frac{1}{2} gt^2.$$

In this case we must write $-h$ for s, since the plane is below the point of projection; we have then

$$\frac{1}{2} gt^2 - u \sin a \cdot t = h;$$

$$\therefore t = \frac{u \sin a \pm \sqrt{2gh + u^2 \sin^2 a}}{g}.$$

Taking the positive value of t and therefore the upper sign of the root in the above expression, we have

$$t = \frac{u \sin a + \sqrt{2gh + u^2 \sin^2 a}}{g},$$

and the horizontal distance from the point of projection of the point at which the particle strikes the plane is

$$u \cos a \cdot t,$$

or

$$\frac{u^2 \sin a \cdot \cos a + u \cos a \sqrt{2gh + u^2 \sin^2 a}}{g}.$$

If it be required to find the direction of the particle's motion when it strikes the plane, we have only to find the vertical and horizontal components of its velocity at the instant. The horizontal component is $u \cos a$, and the vertical component $gt - u \sin a$, that is $\sqrt{2gh + u^2 \sin^2 a}$. Hence the tangent of the angle which the direction of motion makes with the *normal* to the plane is

$$\frac{u \cos a}{\sqrt{2gh + u^2 \sin^2 a}}.$$

193. *A particle is projected with velocity* u *at an elevation* a *; it is required to find the distance from the point of projection of the point where it strikes a plane through*

RANGE ON AN INCLINED PLANE.

the point of projection, inclined at an angle θ to the horizon, and perpendicular to the vertical plane of projection of the particle.

Let BP represent the inclined plane, and P the point

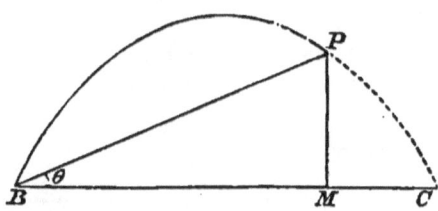

of impact. Let t be the time of flight before reaching P. Draw PM perpendicular to BC. Then

$$BM = u \cos a \cdot t \text{ and } PM = u \sin a \cdot t - \frac{1}{2} g t^2.$$

But since P lies on the plane, $PM = BM \tan \theta$.
Therefore

$$\frac{u \sin a \cdot t - \frac{1}{2} g t^2}{u \cos a \cdot t} = \tan \theta,$$

or $\qquad \tan a - \dfrac{gt}{2u \cos a} = \tan \theta;$

$$\therefore t = (\tan a - \tan \theta) \cdot \frac{2u \cos a}{g}.$$

Hence $BM = u \cos a \cdot t = \dfrac{2u^2 \cos^2 a}{g} \cdot (\tan a - \tan \theta),$

and $\qquad BP = \dfrac{BM}{\cos \theta}$

$$= \frac{2u^2 \cos^2 a}{g \cos \theta} (\tan a - \tan \theta),$$

which gives the range on the inclined plane.

GREATEST RANGE ON AN INCLINED PLANE. 167

This expression for the range we may put into another form, thus:—

$$BP = \frac{u^2}{g \cos \theta} (2 \sin a \cos a - 2 \cos^2 a \tan \theta)$$

$$= \frac{u^2}{g \cos^2 \theta} (\sin 2a \cos \theta - 2 \cos^2 a \sin \theta)$$

$$= \frac{u^2}{g \cos^2 \theta} (\sin 2a \cos \theta - \cos 2a \sin \theta - \sin \theta)$$

$$= \frac{u^2}{g \cos^2 \theta} \{\sin (2a - \theta) - \sin \theta\}.$$

Now if the inclination of the plane, that is θ, be invariable, this expression is greatest when $\sin (2a - \theta)$ is greatest, that is when

$$2a - \theta = \frac{\pi}{2}, \text{ or } a = \frac{1}{2}\left(\frac{\pi}{2} + \theta\right).$$

Hence the greatest range is obtained on an inclined plane when the direction of projection bisects the angle between the plane and the vertical.

194. We have shown that the velocity of a projectile at any point of its path is that which it would gain in falling freely from the directrix of the parabola which it describes, to the point in question. Hence if a number of particles be projected from the same point, in the same vertical plane, and with the same velocity u, all the parabolic paths described by the particles will have the same directrix whose height, h, above the point of projection is given by the equation

$$u^2 = 2gh.$$

Let P be the point of projection, draw PL vertical and of length equal to $\frac{2g}{u^2}$. Through L draw LM horizontal and in the plane of projection. Then LM is the directrix of each of the parabolic paths. Now since the path of each particle passes through P, its focus must be at a distance from P equal to PL, and therefore lie on the circle LKF, whose centre is P and radius PL. This circle is therefore the locus of the foci of all the paths which can be described

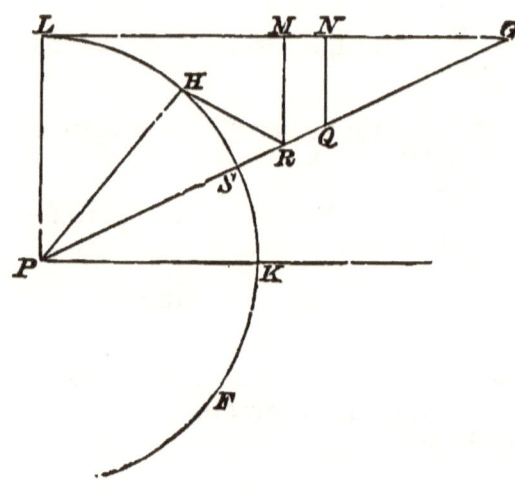

by particles projected from P with a velocity u in the vertical plane containing LM.

195. Suppose we wish to find the range on an inclined plane PG passing through the point of projection. Let H be the focus of the parabolic path. Then if R be the point in PG, which is equidistant from H and the directrix, LM, R will be a point on the parabola, and PR will be the range. Since the tangent to a parabola at any point P bisects the angle between the focal distance of P and the perpendicular from P on the directrix, if H be the focus the direction of projection will bisect the angle HPL.

In order that PR may be a maximum, P, H and R must be in one straight line. Hence if S be the point at which the circle LKF cuts PG, S will be the focus; and if SQ be equal to QN, PQ will be the greatest range which can be obtained on the plane PG, the velocity of projection being u. The direction of projection in this case bisects the angle SPL, that is, the angle between the inclined plane and the vertical, which is the condition for a maximum range which we found in Art. 193.

196. Suppose it required to find the point where the projectile strikes a given plane not passing through the

point of projection. Let P be the point, and PT the direction, of projection, and u the initial velocity. Let

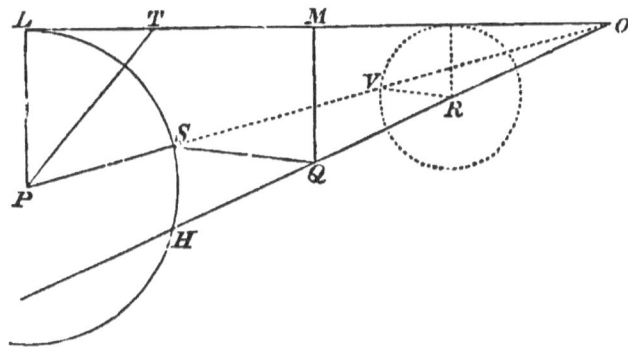

HQ be the line of intersection of the given plane with the vertical plane of projection, which we take as the plane of the paper. Draw PL vertically upwards, and make it equal to $\frac{u^2}{2g}$. Through L draw LM horizontal. Then LM is the directrix of the parabola described by the projectile. With centre P and radius equal to PL describe the circle LSH. Then the focus of the path must lie on this circle. Make the angle TPS equal to the angle LPT. Then S is the focus, and if Q be the point in HQ equidistant from S and the line LM, Q will be a point in the path of the projectile. Hence Q is the point where the particle meets the plane.

Q is of course the centre of the circle which passes through S, touches the line LM, and whose centre lies on HQ. It may be found geometrically by producing LM and HQ till they meet in O, joining OS, in HQ taking any point R, and with R as centre describing a circle touching LO, and cutting SO in V. Then if SQ be drawn parallel to VR and meeting OH in Q, Q will be the point required.

197. In order that the point where the projectile meets the plane HQ may be as far as possible from P, the direction of projection must be such that PSQ is a straight line. For if we suppose any other direction of projection to give a greater range on the plane, let S' be the focus of

the parabola and Q' the point where the projectile meets the plane. Then $S'Q'$ is obviously greater than SQ. But

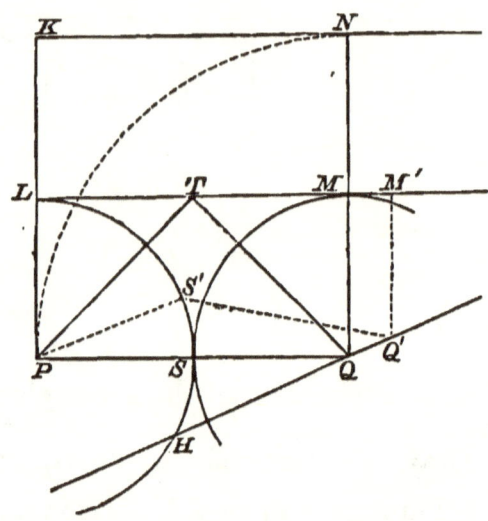

since Q' is a point on a parabola whose focus is S' and directrix LM', $S'Q'$ must be equal to $Q'M'$. Therefore $Q'M'$ is greater than QM, which is obviously not the case. Hence the conditions of the problem will be satisfied when PSQ is a straight line. Q will then be the centre of the circle which touches the line LM and the circle LSH, and has its centre on the line HQ.

To find Q geometrically produce PL to K, making LK equal to PL. Draw KN parallel to LM. By the construction given in the last article find Q the centre of the circle which passes through P, touches KN, and has its centre on the line HQ. Then Q is the point required. For if QN be drawn perpendicular to KN and cutting LM in M, QN is equal to QP. But if S be the point where QP cuts the circle LSH, PS is equal to MN. Hence QM is equal to QS, and Q is the point required.

If PT be drawn bisecting the angle QPL, PT will be a tangent to the parabola at P, and hence PT will be the direction of projection. Also QSP being a focal chord, since tangents at the extremities of a focal chord of a parabola intersect at right angles in the directrix, QT will be the tangent at Q, and will be at right angles to PT. Hence QT is the direction of the particle's motion when it meets the plane HQ, and we see that if the particle be projected so that the point where it meets any plane HQ may be as far as possible from the point of projection, the direction of its motion when it meets the plane is at right angles to the direction of projection.

The curve described by a projectile is frequently called a trajectory.

198. Referring to the figure of the preceding article, since QN is equal to QP, it follows that Q is a point on the parabola whose focus is P and directrix KN. Hence if with P for focus and L for vertex we construct a parabola, the point Q at which this parabola cuts any plane will be the point on that plane farthest from P to which a particle can be thrown with the given initial velocity.

Again, since TQ is at right angles to PT, it follows that QT bisects the angle PQN. Hence QT is the tangent at Q to the parabola whose focus is P and vertex L. But it has been shown that QT is the tangent at Q to the parabola described by the particle projected from P so as to pass through Q. Hence these two parabolas touch at their common point Q. The parabola whose focus is P and vertex L therefore touches each of the parabolas described by particles projected from P, with velocity u, in different directions in the vertical plane of the paper. The various trajectories through P will therefore all lie within the parabola so described, and will each touch it at some point. In such a case the external curve is said to *envelope* the series of inner curves, and is called their *envelope*. The annexed figure represents the series of curves in question.

If we suppose the whole figure to revolve about the line PL, we see that the trajectory of a particle projected from the point P with velocity u in any direction in space lies entirely within, but touches, the paraboloid of revolution generated by the enveloping parabola.

A fountain, proceeding from a rose jet, consists of a

series of particles of water projected with something like the same velocity in different directions, and the general

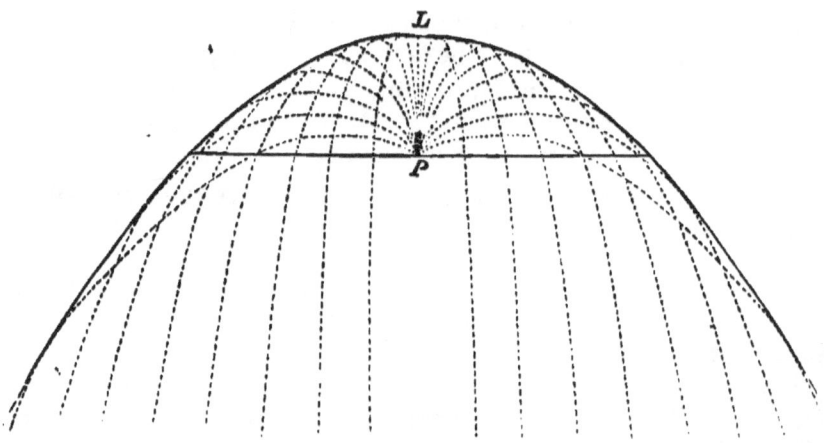

outline of the jet approximates more or less nearly to the above figure. As a rule in fountains the very oblique jets are not present, and the general outline is therefore much narrowed, consisting only of the portion of the figure near the axis. The upper part of the jet however corresponds to the enveloping parabola.

If a bombshell were to burst at P, the lines of the above figure might approximately represent the paths described by the fragments.

199. If a particle of mass m be projected along a smooth plane inclined at an angle θ to the horizon, the resultant force upon it at every instant while in contact with the plane is the resolved part of its weight acting down the plane, that is, $mg \sin \theta$. Hence the motion of the particle will be that of a projectile acted upon always by an acceleration uniform in magnitude and direction, and denoted by $g \sin \theta$. The direction of this acceleration is of course perpendicular to the horizontal line drawn on the plane, and passing through the point of projection. Hence, if u be the initial velocity, and a the angle which the direction of projection makes with the horizontal line through the point of projection lying in the inclined plane, the particle will describe a parabola whose latus rectum is

MOTION OF A PROJECTILE ON A SMOOTH INCLINED PLANE. 173

$\dfrac{2u^2 \cos^2 a}{g \sin \theta}$; the greatest height, measured up the inclined plane, to which the particle will rise, is $\dfrac{u^2 \sin^2 a}{2g \sin \theta}$; the range upon the horizontal plane through the point of projection is $\dfrac{u^2 \sin 2a}{g \sin \theta}$; the directrix is a horizontal line in the inclined plane at a distance $\dfrac{u^2}{2g \sin \theta}$ from the point of projection; the time before the particle reaches the horizontal plane through the point of projection, which may be called the time of flight, is $\dfrac{2u \sin a}{g \sin \theta}$, and the velocity at any point is that due to sliding down the inclined plane from the directrix to the point in question. These expressions may be at once obtained by substituting $g \sin \theta$ for g in the corresponding expressions of Art. 186, the two problems being precisely the same, except that the plane of motion being inclined, in this case, instead of vertical as in Art. 186, the acceleration under which the particle is moving is $g \sin \theta$ instead of g.

200. We have seen that the greatest height, *measured up the inclined plane*, to which the particle attains is $\dfrac{u^2 \sin^2 a}{2g \sin \theta}$. Hence the greatest height, measured *vertically*, to which it rises is $\dfrac{u^2 \sin^2 a}{2g}$, since θ is the inclination of the plane. Hence the particle rises to the same vertical height as if it had been projected from P with velocity u at an elevation a, and its motion had been in a vertical plane.

Again, the distance of the directrix from P, measured up the plane, is $\dfrac{u^2}{2g \sin \theta}$. Hence its vertical height is $\dfrac{u^2}{2g}$, and it follows that the velocity at any point of the parabolic path is that due to falling freely from a plane at height $\dfrac{u^2}{2g}$ above the point of projection. This result is the same as for a particle projected and moving in a vertical plane, and shows that in this, as in other cases, the change of the particle's kinetic energy is simply equivalent to the work done upon it by gravity.

201. In all the above cases of motion, the component of the velocity of the moving point in the direction perpendicular to that of the resultant force remains always constant. Hence the time taken by the particle to move from any one to any other point in its path will be found by drawing straight lines through these points parallel to the direction of the resultant force. Then the numerical measure of the distance between these two lines, divided by the numerical measure of the particle's velocity perpendicular to them, will give the measure of the time of flight from the one point to the other.

202. As examples of the preceding Articles we may take the following.

Ex. 1. *The greatest range of a rifle bullet on level ground is* 20000 *feet. Find its initial velocity; and its maximum range up an incline of* 30°, *neglecting the resistance of the air.*

The range on a horizontal plane is a maximum when the elevation of projection is 45°. The range is then $\frac{u^2}{g}$, if u be the initial velocity. Hence $\frac{u^2}{g} = 20000$;

$$\therefore u^2 = 20000g$$
$$= 640000;$$
$$\therefore u = 800.$$

The velocity of projection must therefore be 800 ft. per second.

Again, it has been shown that the range up an inclined plane through the point of projection is a maximum when the direction of projection bisects the angle between the plane and the vertical, and the range is then represented by the expression

$$\frac{2u^2 \cos^2 a}{g \cos \theta}(\tan a - \tan \theta),$$

where θ is the inclination of the plane, and a the elevation of projection. (See Art. 193.) In the present case $\theta = 30°$, and the range being a maximum, a is therefore equal to 60°, while u is denoted by 800. Hence the expression for the range becomes

$$\frac{2 \cdot (800)^2 \cdot \cos^2 60°}{g \cdot \cos 30°} (\tan 60° - \tan 30°)$$

$$= \frac{2 \cdot 640000}{16 \cdot \sqrt{3} \cdot 4} \left(\sqrt{3} - \frac{1}{\sqrt{3}} \right)$$

$$= \frac{20000 \cdot 2}{3}$$

$$= \frac{40000}{3}$$

$$= 13333 \cdot \dot{3}.$$

Or the range on the inclined plane is $13333\cdot\dot{3}$ feet, that is, two-thirds of the range on the horizontal plane.

203. A simpler way of treating the latter portion of the problem is the following; and it has the advantage of not requiring the general expression for the range on an inclined plane through the point of projection.

Let PQ be the line along which the range is required, the inclination of PQ to the horizon being 30°. Let Q be

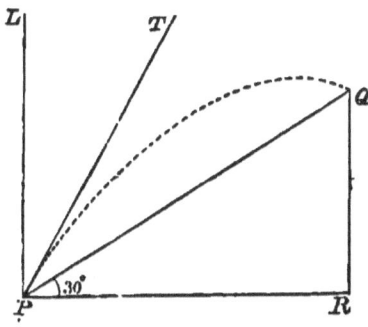

the point where the particle strikes the plane. Then assuming that PT, the direction of projection, bisects the angle QPL, that is, the angle between PQ and the vertical, we have the angle TPR equal to $60°$. Hence the vertical component of the particle's initial velocity is $u \sin 60°$, and the horizontal component, which remains constant, is $u \cos 60°$. Let t be the time of flight from P to Q. Then the height to which the particle will rise in the time t is $u \sin 60° \cdot t - \frac{gt^2}{2}$. Therefore

$$QR = u \sin 60° \cdot t - \frac{1}{2} gt^2.$$

Again, the horizontal velocity being constant and equal to $u \cos 60°$, we have

$$PR = u \cos 60° \cdot t.$$

But since the angle $QPR = 30°$, $QR = PR \tan 30°$;

$$\therefore u \sin 60° \cdot t - \frac{1}{2} gt^2 = u \cos 60° \cdot t \cdot \tan 30°;$$

$$\therefore u \cdot \frac{\sqrt{3}}{2} \cdot t - \frac{1}{2} gt^2 = u \cdot \frac{1}{2} \cdot \frac{1}{\sqrt{3}} \cdot t;$$

$$\therefore 3u - \sqrt{3} \cdot gt = u;$$

$$\therefore t = \frac{2u}{g\sqrt{3}}.$$

And
$$PQ = \frac{PR}{\cos 30°} = \frac{u \cos 60° \cdot t}{\cos 30°}$$

$$= \frac{u}{\sqrt{3}} \cdot \frac{2u}{g\sqrt{3}}$$

$$= \frac{2u^2}{3g} = \frac{2}{3} \, 20000, \text{ as before.}$$

204. **Ex. 2.** *Two balls are projected from the top of a tower, each with a velocity of 50 feet per second, the first at an elevation of 30° and the second at an elevation of 45°. They strike the ground at the same point: find the height of the tower.*

Let AP represent the tower, and let Q be the point where the balls strike the ground. Let the length of AQ

be a feet, and let x feet be the height of the tower. Since the first particle is projected with a velocity of 50 feet per

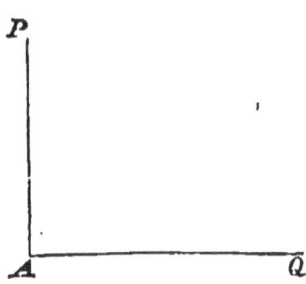

second, at an elevation of 30°, the horizontal component of its velocity is 50 cos 30°, or 25 $\sqrt{3}$ feet, per second. Hence its time of flight from P to Q is $\dfrac{a}{25\sqrt{3}}$ seconds. Now the height to which the particle will rise in t seconds is

$$50 \sin 30° \cdot t - \frac{1}{2} g t^2 \text{ feet} \ldots\ldots\ldots\ldots\ldots(1),$$

and substituting $\dfrac{a}{25\sqrt{3}}$ for t in this expression, we have for the height AP the expression

$$\frac{1}{2} g \cdot \frac{a^2}{25^2 \cdot 3} - 25 \cdot \frac{a}{25\sqrt{3}} \text{ feet}.$$

Again, the horizontal velocity of the second particle is 50 cos 45°, that is $\dfrac{50}{\sqrt{2}}$, feet per second. Hence its time of flight from P to Q is $\dfrac{a\sqrt{2}}{50}$ seconds, and the height above P to which it will rise in this time is found by substituting this value for t in the expression

$$50 \sin 45° \cdot t - \frac{1}{2} g t^2.$$

Hence we obtain for the height AP in feet the expression

$$\frac{1}{2}g \cdot \frac{2a^2}{50^2} - \frac{50}{\sqrt{2}} \cdot \frac{a\sqrt{2}}{50} \quad \ldots\ldots\ldots\ldots\ldots\ldots(2).$$

These two expressions, (1) and (2), for the height of AP must of course be equivalent. Hence, equating them, we have—

$$g\frac{a^2}{50^2} - a = g\frac{a^2}{6.25^2} - \frac{a}{\sqrt{3}},$$

or $\quad g\left(\dfrac{a}{50^2} - \dfrac{a}{6.25^2}\right) = \dfrac{\sqrt{3}-1}{\sqrt{3}};$

$$\therefore a = \frac{\sqrt{3}-1}{\sqrt{3}} \cdot \frac{3.50^2}{g}$$

$$= \frac{2500(3-\sqrt{3})}{g}.$$

Substituting this value of a in either of the above expressions for the height AP, we obtain for this height—

$$\frac{2500}{g}.(3-\sqrt{3})^2 - \frac{2500}{g}(3-\sqrt{3})$$

$$= (2-\sqrt{3}).(3-\sqrt{3})\frac{2500}{g}$$

$$= (9-5\sqrt{3})\frac{2500}{g};$$

and if we take $g = 32$, we obtain for the height of the tower 26·542968... feet.

This last example is useful as indicating how such questions may be solved, without any reference to the parabola, by simply considering the horizontal and vertical motions independently, and finding the time of flight.

205. *Ex.* 3. *A shot is fired at an elevation of* 30° *so as to strike an object at a distance of* 2,500 *feet, and on an*

PROBLEMS ON PROJECTILES.   179

ascent of 1 *in* 40. *Find the velocity of projection, neglecting the resistance of the air.*

This is of course equivalent to finding the velocity of projection in order that the range on a plane through the point of projection inclined to the horizon at the angle $\sin^{-1}\frac{1}{40}$ may be 2500 feet, the direction of the projection being inclined at an angle of 30° to the horizon. We may, therefore, at once use the formula of Art. 193 for the range on an inclined plane, or we may proceed thus:—

Let u be the velocity of projection. Then $\dfrac{u\sqrt{3}}{2}$ is the horizontal velocity of the shot. Now the horizontal distance between the point of projection and the object hit is $2500\dfrac{(40^2-1)^{\frac{1}{2}}}{40}$ feet. Hence the time of flight is—

$$2500\frac{(40^2-1)^{\frac{1}{2}}}{40}\cdot\frac{2}{u\sqrt{3}} \text{ seconds}$$

$$=\frac{125\sqrt{533}}{u} \text{ seconds.}$$

Now the vertical velocity of the shot, initially, is $u \sin 30°$, that is, $\dfrac{u}{2}$. Hence the height to which the shot will ascend in the above time is

$$\frac{u}{2}\cdot\frac{125\sqrt{533}}{u}-\frac{1}{2}g\cdot\frac{125^2.533}{u^2} \text{ feet}$$

$$=\frac{125}{2}\sqrt{533}-\frac{1}{2}g\frac{125^2.533}{u^2} \text{ feet.}$$

But the height of the object above the point of projection is $\dfrac{2500}{40}$ feet, since it lies on an incline of 1 in 40. Hence we must have—

$$\frac{125}{2}\cdot\sqrt{533}-\frac{1}{2}g\cdot\frac{125^2.533}{u^2}=\frac{2500}{40};$$

$$\therefore u^2=\frac{g}{2}\cdot125^2.533\cdot\frac{40}{2500(\sqrt{533}-1)},$$

whence $u = 310\cdot6\ldots$, or the velocity with which the shot must be fired is $310\cdot6\ldots$ feet per second, g being taken equal to 32.

EXAMINATION ON CHAPTER III.

1. What is the greatest height to which a particle will rise if projected at an elevation of 30° with a velocity equal to that which it would gain by falling freely through a vertical height of 100 feet?

2. Show that, at any time during its flight, the kinetic energy of a particle projected with given velocity depends only on the vertical distance above or below the point of projection, and is independent of the elevation of projection.

3. Find the range in vacuo of a rifle-bullet projected with a velocity of 1200 feet per second, the direction of projection making an angle $\sin^{-1}\frac{1}{10}$ with the horizon.

4. A particle is projected at an elevation of 60° with a velocity of 200 feet per second. Find the latus rectum of the parabola which it describes, and the height of the focus above the horizontal plane through the point of projection.

5. A number of particles are projected from a point with different velocities and in different vertical planes but at the same elevation. Show that the foci of their subsequent paths all lie on a right circular cone.

6. Find the direction in which a stone must be projected with a velocity of 80 feet per second in order to strike a small bird on the top of a vertical pole 20 feet higher than the point of projection, and 30 feet in front of it.

7. A particle is projected from the lowest point of a smooth inclined plane, 40 feet in length, directly up the plane, with a velocity of 50 feet per second, and, passing over the edge of the plane, describes a portion of a parabola. If the inclination of the plane to the horizon be

45°, find the point where the particle will strike the horizontal plane through the point of projection.

8. A particle is projected along a smooth plane, inclined 30° to the horizon, with a velocity of 100 feet per second, and meets the horizontal plane through the point of projection at a distance of 625 feet from that point. Find the direction of projection.

9. A train is moving at the rate of 30 miles an hour, and a person in one of the carriages projects a ball vertically upwards with a velocity of 8 feet per second. Find the latus rectum of the parabola which the ball will describe.

10. The earth's equatorial radius being 3962·5 miles, if a particle be allowed to fall from the top of a mast at the height of 100 feet above the deck of a ship, find where it will meet the deck.

EXAMPLES ON CHAPTER III.

1. Determine the direction and velocity of projection that a shot may strike the ground after 16 seconds at a distance of 9000 feet from the point of projection.

2. A plane is inclined at an angle of 30° to the horizon: a particle is projected from a point in it in a direction making an angle of 60° with the horizon. Compare the ranges on the plane when the particle is projected up the plane, and when projected down it.

3. A particle is projected with a velocity 128 feet per second at an elevation of 60°: find the direction of its motion when at a height of 128 feet, and its distance from the point of projection at that time.

4. Given the time of flight of a particle on a horizontal plane, find the greatest height to which it rises.

5. A body is projected horizontally, with a velocity of 128 feet per second, from a point whose height above the ground is 512 feet: find the direction of its motion (1)

when it has fallen half way to the ground, (2) when half the whole time of falling has elapsed.

6. A shot is fired with a velocity of 1000 feet per second against a tower, whose height is 100 feet, and horizontal distance from the gun one half the greatest range which can be obtained with the given velocity of projection. Find between what limits the angle of projection must lie in order that the shot may hit the tower, neglecting the resistance of the air.

7. Heavy particles slide down chords of a vertical circle from rest at the highest point. Show that the locus of the foci of the parabolic paths they describe after leaving the chords is a circle, and that of their vertices is an ellipse whose axes are in the ratio of 2 to 1.

8. Two particles are projected simultaneously from the same point in directions which are in the same vertical plane; show that, if they impinge on each other afterwards, their paths must coincide.

9. If particles be projected from a point with the same velocity, the foci of the parabolas described lie on the surface of a sphere.

10. Particles are projected from the same point and in the same plane so as to describe equal parabolas: show that the vertices of their paths lie on a parabola.

11. A heavy particle is projected from a given point in a given direction so as to touch a given straight line. Give a geometrical construction for determining the point of contact, and the elements of the trajectory.

If the direction of projection be not fixed, find the trajectory, so that the velocity of projection may be the least possible.

12. A particle is projected with given velocity: find the elevation of projection that its direction of motion may be horizontal when it strikes a given inclined plane through the point of projection.

13. A ball is projected horizontally from the top of a

EXAMPLES. 183

staircase, each stair of which is a feet high, and c feet broad, with a velocity represented by $\sqrt{2gnc}$; find from which step it will first rebound.

14. From a point in a fixed horizontal line, a particle is projected, with a given velocity, along an inclined plane passing through the line, so that its horizontal range may be the greatest possible. If the inclination of the plane is varied, the locus of the vertex of the parabolic path will be a hyperbola.

15. If τ be the time in which a projectile passes from any point P to the vertex of its parabolic path, and t the time of falling, under the action of gravity only, from rest at the vertex to the focus, show that $\tau^2 + t^2$ varies as the distance of P from the directrix.

16. A particle is projected at right angles to an inclined plane with the velocity which would be acquired in falling freely through a space equal to $\dfrac{3}{8}$ of the range on the plane: find the inclination of the plane.

17. From a point on an inclined plane two particles are projected with the same velocity in the same vertical plane in directions at right angles to each other; show that the difference of their ranges is constant.

18. If a body be projected at an angle a to the horizon with a velocity equal to 32 feet per second, its direction of motion is inclined at an angle $\dfrac{a}{2}$ to the horizon at the end of the time $\tan\dfrac{a}{2}$, and at the angle $\dfrac{\pi-a}{2}$ at the end of the time $\cot\dfrac{a}{2}$.

19. ABC is a right-angled triangle in a vertical plane with the hypotenuse AB horizontal. A particle projected from A passes through C and falls at B; show that the angle of projection is $\tan^{-1}(2 \operatorname{cosec} 2A)$, and that the latus rectum of the path described is equal to the height of the triangle.

20. If a particle impinge perpendicularly on a plane through the point of projection inclined at an $<a$ to the horizon, show that its range on the plane is equal to—
$$\frac{2v^2}{g} \cdot \frac{\sin a}{1 + 3 \sin^2 a},$$
where v is the velocity of projection.

21. A particle is attracted to one centre of force and repelled by another, both forces varying as the distance: show that, if the absolute intensities of the forces are equal, the path of the particle is a parabola.

22. Heavy particles are projected horizontally with different velocities from the same point: show that the extremities of the latera recta of the parabolas which they severally describe lie on a cone, of which the axis is vertical, and the vertical angle $2 \tan^{-1} 2$.

23. A smooth rectangle, the length of whose edges are respectively 20 and 10 feet, is placed with its longer edge horizontal and its plane inclined 30° to the horizon: find the velocity with which a particle must be projected from one corner so as to leave the plane horizontally at the opposite corner, and show that the horizontal range after leaving the plane is one-half that described on the plane.

24. A heavy particle is projected from a point so as to pass through another point not in the same horizontal line with it. Show that the locus of the focus of its path will be a hyperbola.

25. A particle slides from rest down a smooth inclined plane: show that the distance from the foot of the plane of the focus of the parobola which the particle describes after leaving the plane is equal to the height of the plane.

26. The parabolic paths of two projectiles have the same focus; if tangents be drawn to the parabolas from any point in their common axis, the velocities of the projectiles at the points of contact are equal.

27. If t and t' be the two times of flight on an inclined plane through the point of projection, corresponding to

EXAMPLES. 185

any given range short of the greatest, and a the inclination of the plane, prove that—

$$t^2 + t'^2 + 2tt' \sin a$$

is independent of a, the velocity of projection being given.

28. How must a ball be projected from a point distant a feet from a vertical wall which is c feet high, so that it may just pass over the wall at an angle of $45°$ with the horizon, and fall at a distance of b feet from the wall?

29. The time during which a projectile moves from one end to the other end of a focal chord is equal to the time in which it falls vertically from rest through a space equal to the length of the chord.

30. If a particle be projected, at an elevation of 60 degrees, up a plane inclined at an angle of 30 degrees to the horizon and passing through the point of projection, show that the range in feet is 16 times the square of the time of flight in seconds.

31. Particles are placed along a rod which revolves about a hinge in a vertical plane; when ascending and inclined at an angle of $45°$ to the vertical, the rod is suddenly stopped, and the particles proceed to move freely. Show that they will all be in the vertical line through the hinge at the same instant; and that if one particle strikes the hinge at the same moment as another reaches the highest point of its path, the original distances of these particles from the hinge were as 1 to 4.

32. Given the point of projection of a projectile and the range on a horizontal plane, find a geometrical construction for the focus and directrix of the path, in order that it may pass through a given point.

33. Two shots are fired at a tower in parallel directions from two points in the horizontal plane through its base, distant a and b respectively from the tower. They both hit the top of the tower. Prove that, if the initial velocity be the same in each case, the height of

the tower is $\dfrac{ab}{a+b}\tan a$, where a is the inclination to the horizon of the direction in which the shots are fired.

34. Two inclined planes intersect in a horizontal line, their inclinations to the horizon being a, β: if a particle be projected from a point in the former at right angles to it so as to strike the latter at right angles, the velocity of projection must be—

$$\sin\beta\sqrt{\dfrac{2ga}{\sin a - \sin\beta\cos(a+\beta)}},$$

a being the distance of the point of projection from the intersection of the planes.

35. A particle is projected with a given velocity in a given direction. After what time will it be moving at right angles to this direction, and what will be its velocity and position at that time?

36. Give a geometrical construction for determining the two directions in which a particle may be projected from a given point A with a given velocity to pass through another given point B.

If the velocity of projection be $\sqrt{2ga}$, and if h, k be the vertical and horizontal distances of B and A, show that when these directions coincide,—

$$k^2 = 4a(a-h).$$

37. On the moon there seems to be no atmosphere, and gravity is about one-sixth of that here on earth. What space of country would be commanded by the guns of a lunar fort able to project shot at 1600 feet per second?

38. Show that the whole area commanded by a gun on a hill-side is an ellipse whose focus is at the gun, whose eccentricity is the sine of the inclination of the hill to the horizon, and whose semi-latus rectum is twice the greatest height to which the gun could send a ball.

39. Particles are projected from the same point with equal velocities; prove that the vertices of their paths lie

EXAMPLES. 187

on an ellipse. If they be all equally elastic, and impinge on a vertical wall, prove that the vertices of their paths after impact lie on an ellipse.

40. A shot of m pounds is fired from a gun of M pounds, placed on a smooth horizontal plane and elevated at an angle a. Prove that, if the muzzle velocity of the shot relative to the ground be V, the range will be—

$$2\frac{V^2}{g} \cdot \frac{\left(1 + \frac{m}{M}\right)\tan a}{1 + \left(1 + \frac{m}{M}\right)^2 \tan^2 a}.$$

41. A particle is projected from a platform with velocity V and elevation β. On the platform is a telescope, fixed at elevation a. The platform moves horizontally in the plane of the particle's motion, so as to keep the particle always in the centre of the field of view of the telescope. Show that the original velocity of the platform must be $V \frac{\sin(a-\beta)}{\sin a}$, and its acceleration $g \cot a$.

CHAPTER IV.

ON COLLISION.

206. IF a particle of mass m be moving with a velocity v, and be retarded by a constant force which brings it to rest in time t, then the measure of this force we have seen to be $\frac{mv}{t}$. In fact the primary notion of momentum is the effect, or *product*, of a force acting during a finite time upon matter free to move, and it follows from the second law of motion that that which is produced by a given force acting for a given time on any quantity of matter free to move, is always the same amount of momentum, and this is proportional to the algebraical product of the force and the time during which it acts. Now, suppose the time t during which the particle is brought to rest to be made very small. Then the force required to bring it to rest is very large, and if we suppose t so small that we are unable to measure it, then the force becomes very great, but we are unable to obtain its measure. In this case, then, we are compelled to adopt some other mode of considering the question. Since we are unable to measure the time during which the velocity of the particle is being destroyed, we leave the element of time out of consideration altogether; the force we call an impulsive force, and we measure its impulse by the whole momentum which it destroys.

Hence it will be seen that the nature of an impulse is totally different from that of a force, and the two things cannot be compared, for an impulse is the same as the ultimate effect of a finite force acting for a finite time. Indeed, we speak of the impulse instead of the force itself

DIRECT AND OBLIQUE IMPACT. 189

simply on account of our inability to measure very short intervals of time, and to observe what takes place during them, there being no case in nature in which a finite change of motion is produced in an indefinitely short time; for, whenever a finite velocity is generated or destroyed in nature, a finite time is occupied in the process, though we are frequently unable to measure it even approximately. For example, if two balls strike one another, each of them will be more or less compressed or indented, and they will remain in contact for a finite time; though the harder the balls, other things being the same, the less will they be compressed, and the shorter will be the time during which they will remain in contact. We may here notice that some bodies, when they become indented by impact, retain the indentation, while others more or less completely resume their original form. The first class are generally called *inelastic*, and the second class are called *elastic* bodies.

207. DEF. Two bodies are said to impinge *directly* upon one another when the surface of either at the point of contact is perpendicular to the direction of their relative velocity.

When this condition is not fulfilled, the impact is said to be *oblique*.

From the definition of direct impact it follows that when two bodies impinge directly upon one another, the mutual action between them is entirely in the same straight line as the velocity of one relative to the other.

Newton found that if he allowed two bodies to impinge directly upon one another, their relative velocity after impact bore a constant ratio to that before impact, so long as the materials of which the bodies were composed were unchanged, but was in the opposite direction. The numerical value of this ratio is called the coefficient of mutual elasticity of the two substances, and is generally denoted by e. (See Art. 251.)

If the incidence be oblique but the surfaces of the bodies smooth, then the whole action between the two is in the direction of the common normal to the surfaces at the point of incidence. In this case the component of the relative velocity of the bodies in the direction of the common normal is changed by impact in the ratio of 1 to e, and reversed in direction, while the component of the relative velocity in a direction perpendicular to this remains unchanged.

If the surfaces of the bodies be rough, there is also a tangential action between them, and the change in the relative velocity is not wholly in the direction of the common normal to the surface at the point of contact.

When e is equal to 1, or the velocity of one body relative to the other is the same after impact as before, but in the opposite direction, the bodies are said to be *perfectly elastic*. If e be equal to 0, or the bodies after impact go on moving together, they are said to be *inelastic*. No bodies occurring in nature are either perfectly elastic or inelastic. Glass and some crystals are amongst the most elastic bodies known, while clay, putty, etc., have very little elasticity.

208. If the two bodies are of the same material, the numerical value of the ratio of their relative velocities after and before impact is called the *coefficient of elasticity* of the particular material of which they are composed.

It should be observed that the coefficient of elasticity is a very different quantity from the *elasticity* or *modulus of elasticity* of a substance, as those terms are employed in connection with the strength of materials. The elasticity or modulus of elasticity is properly defined as the limiting ratio of $\frac{\text{stress}}{\text{strain}}$ when each is indefinitely diminished, and for every kind of strain which may be produced this ratio gives a corresponding modulus of elasticity. The coefficient of elasticity, as defined above, is quite independent of any modulus of this sort. For example, the coefficient of elasticity, as measured by ratio of velocity after and before impact, may be much greater for india-rubber than for wrought iron, but the ratio $\frac{\text{stress}}{\text{strain}}$ is *enormously* greater in the case of wrought iron than in the case of india-

IMPACT ON A MOVING PLANE. 191

rubber for every kind of strain except cubic compression. For a further discussion of the coefficient of elasticity see Arts. 235-241.

We may remark that the impact of two spheres is direct when the line joining their centres at the moment of impact is in the direction of their relative motion.

We shall in this chapter consider the impact of particles and spheres against one another and against planes only, and, except when the contrary is stated, we shall suppose their surfaces smooth.

209. The simplest case of impact is that of a particle impinging directly upon a fixed inelastic plane. In this case, since the coefficient of mutual elasticity is zero, the particle will come to rest. Now, if m be the mass of the particle and v its velocity, its momentum will be denoted by mv, and as this is entirely destroyed by the impact, the measure of the impulse which the plane exerts on the particle is mv; but since action and reaction are equal and opposite, this is also the measure of the impulse of the pressure exerted by the particle on the plane.

210. Next, suppose a particle of mass m, and moving with velocity v, to impinge directly on a fixed plane, the coefficient of elasticity being e. In this case the particle will rebound from the plane with velocity ev. The impulse exerted by the plane on the particle must therefore destroy the velocity v with which it is moving, and generate a velocity ev in the opposite direction. Hence the whole change of the particle's velocity is numerically equal to $v + ev$, and the whole change in its momentum to $mv(1+e)$. Hence the measure of the impulse between the particle and plane is $mv(1+e)$.

If the plane be perfectly elastic we have e equal to 1, and the impulse is measured by $2mv$.

211. Next, suppose the plane on which the particle impinges to be moving with a velocity V in the same direction as the particle before impact. Then the velocity of the particle relative to the plane is $v - V$ before impact, and after impact it is $e(v - V)$, but in the opposite direction. If we adopt the usual convention with respect to

sign, we may denote the velocity of the particle relative to the plane after impact by $-e(v-V)$. The whole change of the velocity of the particle is $(v-V)(1+e)$, and the change of its momentum is $m(v-V)(1+e)$, without regard to sign. This last expression is the measure of the whole impulse between the particle and the plane.

Since the velocity of the particle before impact was v, and the change of velocity produced by the impact is $(v-V)(1+e)$ in a direction opposite to that of v, it follows that the velocity of the particle after impact is $v-(v-V)(1+e)$, and is in the same direction as before, or in the opposite direction, according as the sign of this expression is positive or negative.

If e be zero, or the plane inelastic, the velocity after impact is V, as of course it should be. If e be unity or the elasticity perfect, the velocity after impact is $2V-v$, and this is in the same direction as before, or in the opposite direction, according as V is greater or less than $\dfrac{v}{2}$.

The above investigations are true for spherical balls as well as particles, provided that their centres of gravity coincide with their centres of figure, and that they have no motion of rotation unless their surfaces be perfectly smooth. If the spheres be rough, and they have a motion of rotation, or if their centres of gravity do not coincide with their centres of figure, the problem becomes much more complicated, and requires the principles of Rigid Dynamics, to which subject all problems on the motion of spheres or of any rigid bodies of finite dimensions properly belong.

212. Suppose a particle of mass m, moving with velocity v along the line QP, to impinge obliquely at P, upon the smooth plane AB, the coefficient of elasticity between the particle and plane being e. It is required to find the motion of the particle immediately after impact, and the impulse on the plane. Let PT be the direction of motion after impact, PN the normal at P to the plane. Let the angle QPN be denoted by a, and the angle TPN by θ. Then the velocity of the particle

before impact may be resolved into two components; viz., $v \sin a$ along the plane, and $v \cos a$ perpendicular to the

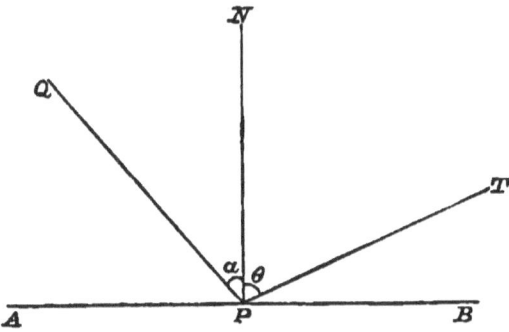

plane. Since the plane is smooth, it is only the latter component which is altered by the impact, and this is replaced by a velocity $ev \cos a$ in the opposite direction. The velocity after impact is therefore the resultant of the two velocities, $v \sin a$ along the plane and $ev \cos a$ perpendicular to the plane, and is therefore numerically represented by—

$$v \sqrt{\sin^2 a + e^2 \cos^2 a};$$

and if PT be the direction of motion after impact,—

$$\cot TPN = \frac{ev \cos a}{v \sin a},$$

or $\quad \cot \theta = e \cot a.$

If the elasticity be perfect, or e equal to 1, we have θ equal to a, or the angle of reflection equal to the angle of incidence.

It remains to determine the impulse of the pressure between the particle and the plane. The component of the velocity of the particle perpendicular to the plane before impact is $v \cos a$, and after impact it is $ev \cos a$ in the opposite direction. The whole change of velocity produced in the particle by the impulse is therefore $v \cos a\,(1 + e)$, and the change of momentum is measured by $mv \cos a\,(1 + e)$, which is therefore the measure of the impulse.

213. If a particle impinge obliquely on a *rough* plane

whose coefficient of friction is μ, then, besides the impulsive pressure, there will be an impulsive friction called into play, which will be at each instant proportional to the pressure, and such that its effect is to diminish the velocity of the particle parallel to the plane.

Referring to the figure of the preceding Article, since the velocity of the particle perpendicular to the plane is reversed in direction and diminished in the ratio of 1 to e by the impact, it follows, as before, that the measure of the impulse is $mv \cos a\,(1+e)$. Now, the impulse of the friction called into play is μ times this, that is—

$$\mu \cdot mv \cos a\,(1+e),$$

and diminishes the particle's velocity parallel to the plane. Hence the velocity, parallel to the plane, of the particle after impact is $v \sin a - \mu v \cos a\,(1+e)$. Therefore, if, as in the preceding example, θ denote the angle the direction of motion after reflection makes with the normal, we have—

$$\cot \theta = \frac{ev \cos a}{v \sin a - \mu v \cos a\,(1+e)}$$
$$= \frac{e \cos a}{\sin a - \mu \cos a\,(1+e)},$$

which gives the direction of motion after impact, and the velocity can be found as in the previous case. (See also Art. 222, p. 203.)

If the surface upon which the particle impinges be curved, the effect of the impact is the same as if the surface were replaced by its tangent plane at the point at which the particle strikes it. Thus in the preceding Article, if PN be the normal at P to the surface AB, the whole of the reasoning will be equally true whether the surface AB be plane or curved.

214. Hitherto we have considered the obstacle against which the moving particle impinges to be either fixed, or made to move in such a manner that its velocity is unaffected by the collision. We proceed now to the consideration of the collision of two particles, or two spheres, each free to move. It will be seen that in order that two spheres may impinge *directly* upon one another, the line

DIRECT IMPACT OF TWO BALLS. 195

joining their centres at the time of impact must be the line of motion of either *relative to the other*.

Suppose two particles, or spheres, whose masses are respectively M and m, to be moving in the same direction with velocities V and v respectively, of which V is the greater, and to impinge directly upon each other; it is required to find the motion of each after impact, the coefficient of elasticity being e.

Let V_1, v_1 be their respective velocities after impact. Then whatever be the impulsive action between them at the moment of impact, the impulse on the first must be equal and opposite to that exerted on the second. Hence the momentum generated in the first must be numerically equal but opposite in direction to that generated in the second. Hence, whatever momentum reckoned in the positive direction may be lost by the first, the same amount must be gained by the second, and *vice versâ;* consequently the algebraical sum of the momenta of the two balls must be the same after impact as before. Therefore

$$MV_1 + mv_1 = MV + mv \quad \ldots\ldots\ldots\ldots\ldots\ldots(\text{I.}).$$

Again, the relative velocity after impact is to that before impact as $-e$ to 1, and the velocity of the first relative to the second before impact is $V-v$, and after impact it is $V_1 - v_1$. Therefore—

$$V_1 - v_1 = -e(V - v) \ldots\ldots\ldots\ldots\ldots\ldots(\text{II.}).$$

These two equations, (I.) and (II.) determine V_1 and v_1.

From (II.) we have—

$$V_1 = v_1 - e(V - v) \ldots\ldots\ldots\ldots\ldots(\text{III.}).$$

Substituting in (I.) we get—

$$(M + m)v_1 = MV + mv + eM(V - v);$$

$$\therefore v_1 = \frac{MV + mv + eM(V - v)}{M + m} \ldots\ldots\ldots\ldots(\text{IV.}).$$

Hence from (III.)

$$V_1 = \frac{MV + mv - em(V - v)}{M + m} \ldots\ldots\ldots\ldots(\text{V.}).$$

If the balls be inelastic, e is zero, and from equation (II.) we have V_1 equal to v_1, or they proceed with a common velocity; in other words, they do not separate. Substituting in equation (I.) we have, in this particular case,—

$$V_1 = v_1 = \frac{MV + mv}{M + m} \quad \dots\dots\dots\dots\dots(VI.).$$

This of course follows immediately from the general expressions given above for V_1 and v_1, by putting e equal to 0 in them.

215. If the balls before impact are moving in opposite directions, we have merely to give opposite signs to V and v.

It remains to determine the impulse of the pressure between the two balls. We have determined the velocity of each after impact on the assumption that the changes of their respective momenta are equal and opposite. To find the impulse, we need only consider the change of momentum of one of the balls.

The velocity of the first before impact is V, and after impact its velocity is V_1, hence the change of its momentum is $M(V - V_1)$, that is—

$$M\left\{V - \frac{MV + mv - em(V-v)}{M + m}\right\}.$$

Hence if I denote the impulse between the balls, we have—

$$I = M\left\{V - \frac{MV + mv - em(V-v)}{M + m}\right\}$$

$$= \frac{Mm}{M+m}(V-v)(1+e)\dots\dots\dots\dots\dots(VII.).$$

We may notice that if the masses of the balls are equal and the elasticity perfect, or e equal to 1, it follows from equations (IV.) and (V.) that $v_1 = V$ and $V_1 = v$, or the balls exchange their velocities. If in the equations we make m infinite, we obtain the same result as in Art. 211; in fact, we revert to the case of impact against a moving obstacle whose velocity is unchanged by the collision.

OBLIQUE IMPACT OF TWO SPHERES.

216. We proceed now to consider the case of the impact of two smooth spheres not moving in the same or in opposite directions. We shall investigate only the case in which the centres of the two spheres are moving in the same plane. The solution of the general case is precisely similar, but the geometry is more difficult.

Let the two spheres be called A and B respectively, M denoting the mass of A, and m that of B.

Let KG be the straight line drawn through O, O' the centres of the spheres at the moment of impact. Let DO

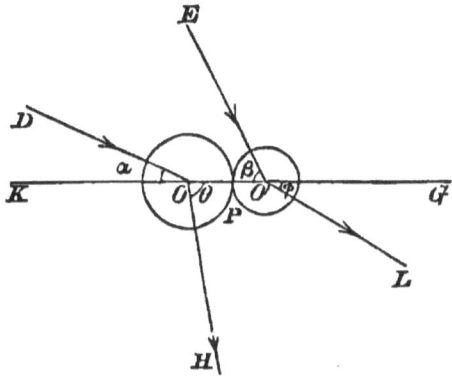

be the direction in which the sphere A is moving before impact, and V its velocity, EO' the direction of motion of B before impact, and v its velocity. Let V_1, v_1 be their respective velocities after impact, and OH, $O'L$ the directions in which they are respectively moving. Let the angles DOK, $EO'K$, HOG, $LO'G$ be denoted by a, β, θ, ϕ respectively. Let e be the coefficient of mutual elasticity.

The velocity of A before impact may be resolved into two components, $V \cos a$ along OO' and $V \sin a$ perpendicular to OO', while that of B may be resolved into $v \cos \beta$ and $v \sin \beta$ in the same directions respectively. Now, since the spheres are smooth, the whole action between them is in the line OO. Hence the components of their velocities perpendicular to this line remain unaffected. We have therefore—

$$\left. \begin{array}{l} V \sin a = V_1 \sin \theta \\ v \sin \beta = v_1 \sin \phi \end{array} \right\} \quad \ldots\ldots\ldots\ldots\ldots\ldots (\text{I.}).$$

Since action and reaction are equal and opposite, the impulse upon A is equal and opposite to that upon B; hence the change of A's momentum must be equal and opposite to that of the momentum of B, and the change in each takes place wholly in the direction of the impulse; that is, along the line OO'. Hence, equating the momentum lost by A to that gained by B, we have—

$$M(V \cos a - V_1 \cos \theta) = m(v_1 \cos \phi - v \cos \beta),$$

or $\quad MV_1 \cos \theta + mv_1 \cos \phi = MV \cos a + mv \cos \beta$......(II.).

Also the velocity, in the direction of the impulse, of either sphere relative to the other after impact is to that before impact as $-e$ to 1. Hence

$$V_1 \cos \theta - v_1 \cos \phi = -e(V \cos a - v \cos \beta)......\text{(III.)}.$$

The four equations (I.), (II.), and (III.) completely determine V_1, v_1, θ and ϕ in terms of the given quantities.

From equations (II.) and (III.) we obtain, precisely as in the preceding Article,—

$$V_1 \cos \theta = \frac{MV \cos a + mv \cos \beta - em(V \cos a - v \cos \beta)}{M + m} \quad \text{..(IV.).}$$

$$v_1 \cos \phi = \frac{MV \cos a + mv \cos \beta + eM(V \cos a - v \cos \beta)}{M + m} \quad \text{..(V.).}$$

Also from (I.)—

$$V_1 \sin \theta = V \sin a,$$

and $\quad v_1 \sin \phi = v \sin \beta,$

hence the components of the velocities of A and B along and perpendicular to OO' are known, and the values of V_1 and v_1 can be at once written down. Again, from (I.) and (IV.),—

$$\tan \theta = \frac{V \sin a (M + m)}{VM \cos a + mv \cos \beta - em(V \cos a - v \cos \beta)} \quad \text{.....(VI.),}$$

$$\tan \phi = \frac{v \sin \beta (M + m)}{MV \cos a + mv \cos \beta + eM(V \cos a - v \cos \beta)} \quad \text{...(VII.).}$$

Hence the direction of motion of each sphere, after impact is found.

To find the impulse of the pressure between the balls, we observe that the momentum of A in the direction OO' before the impact is $MV\cos a$, and this is changed by the pressure into $MV_1 \cos\theta$. The measure of the impulse is therefore—

$$M(V\cos a - V_1 \cos\theta);$$

or, if this impulse be denoted by I, we have—

$$I = M\left\{V\cos a - \frac{MV\cos a + mv\cos\beta - em(V\cos a - v\cos\beta)}{M+m}\right\}$$

$$= \frac{Mm}{M+m}(V\cos a - v\cos\beta)(1+e) \dots\dots\dots\dots\dots(\text{VIII.}).$$

217. In all the cases we have investigated the expression for the impulse involves the factor $1+e$, and is therefore greater in the ratio of $1+e$ to 1 than it would have been had e been zero, but all other circumstances the same. Thus if, in any case of collision, I' measure the impulse when the coefficient of elasticity is zero, and I be the measure of the impulse when, all other things being the same, the coefficient of elasticity is e, we have—

$$I = I'(+e).$$

Now, if we examine somewhat more closely into what takes place when two bodies strike one another, we find that, in the first place, each of them becomes compressed or indented; but if they are elastic, they subsequently recover more or less completely their original form. If they are inelastic, they remain indented, and move on together with a common velocity. Now, up to the instant of greatest compression the action is the same whether the balls are elastic or inelastic, and therefore, *at* this instant, even though the balls be elastic, they will be moving with a common velocity, and the change of momentum produced in either ball up to this instant will be the same as though they were inelastic. This change of momentum is sometimes improperly called the "*force of compression.*" (See Art. 239.) We have denoted it by I'. Now, in the case of elastic balls, after the compression, or indentation, has attained its maximum, the balls begin to recover their form. The parts which have been compressed consequently swell out against one another, and the force which they exert on one another serves to separate the balls. The impulse of this force is sometimes improperly called the "*force of restitution.*" The time taken by the balls to recover their form, and therefore the time during which this force acts, is so short, that we are unable to measure it, and we are consequently compelled, in respect of the force of restitution, like that of compression, to consider the whole momentum generated; in other words, to consider the impulse. Let this impulse be denoted by I''. Then the whole change of momentum produced in either ball during the impact is that due to the force of compression, together with that due to the force of restitution, and is therefore numerically equal to $I' + I''$. Hence—

$$I = I' + I''.$$

But $\qquad I = I'(1 + e);$

therefore $\qquad I'' = eI',$

or the "force of restitution" is equal to e times the "force of compression." If e be equal to unity, or the elasticity be perfect, the "forces of restitution and compression" are equal.

218. In some treatises the coefficient of elasticity is defined as the ratio of the "force of restitution" to that of compression, and it is stated as the result of experiment that this is constant for the same materials. It should, however, be borne in mind that the element observed in experiments on this subject is not the forces which act during the collision, but the velocities of the balls before and after impact, and the measures of the forces of compression and restitution are subsequently deduced from the results of these observations by the Second Law of Motion. It would therefore seem that the method adopted in the preceding Articles is the more natural way of treating the subject.

The coefficient of elasticity is sometimes called "*the coefficient of restitution.*"

219. If when the two bodies impinge upon one another

EXAMPLES ON COLLISION. 201

they be acted upon by some "finite" force, as, for example, gravity, then, since the time during which they remain in contact is so short that we cannot measure it, the effect produced by the finite force in that time will also be immeasurably small. We may therefore neglect it altogether while considering what happens during the collision. The subsequent motion of the bodies will, however, of course depend upon the forces which act upon them.

220. As illustrations of the preceding Articles, we will consider a few examples.

Ex. 1. *A ball of 8 pounds, moving with a velocity of 12 feet per second, strikes directly a ball of 12 pounds, moving with a velocity of 8 feet per second in the opposite direction, the coefficient of elasticity being* $\frac{1}{2}$. *Find the velocity of each after impact, and the impulse of the pressure between the two.*

Let V denote the velocity of the first ball in feet per second, v that of the second, after impact. We will consider velocity positive when in the direction in which the first ball is moving before impact. Now, since the impulses on the two balls are equal and opposite, the momentum gained by the one is equal to that lost by the other; hence the sum of the momenta of the two is the same after impact as before. Now, one pound being taken as the unit of mass, and a velocity of a foot per second as the unit of velocity, the momentum of the first ball before impact will be represented by 8×12, and that of the second by -12×8. Therefore—

$$8 \cdot V + 12v = (8 \times 12) - (12 \times 8) = 0\ldots\ldots\ldots\ldots(\text{I.}).$$

Again, the velocity before impact of the first ball relative to the second is $12 + 8$ feet per second, while after impact it is $V - v$. Therefore, since the coefficient of elasticity is $\frac{1}{2}$, we have—

$$V - v = -\frac{1}{2}(12 + 8)$$

The equations (I.) and (II.) determine V and v.

From (I.)—
$$v = -\frac{2}{3}V.$$

Therefore, from (II.),—
$$\frac{5}{3}V = -10,$$

or $V = -6$.

Hence $v = 4$,

and thus the velocities after impact are found. In this particular case we see that the velocity of each is reversed in direction and its measure reduced by one half.

To find the impulse, we observe that the momentum of the first ball before impact was represented by 8×12, while its momentum after impact is $8 \times (-6)$. Hence the change of momentum produced by the impact is represented by 8×18 or 144; that is, the impulse is equivalent to a velocity of 144 feet per second generated in one pound of matter.

The same results might of course be obtained from the formulæ of Art. 214.

221. *Ex. 2. A ball falls from rest at a height of 20 feet above a fixed horizontal plane. Find the height to which it will rebound, the coefficient of elasticity being $\frac{3}{4}$ and the value of g being 32 foot-second units.*

Let the velocity of the ball, when it strikes the plane, be denoted by v. Then, since it has fallen from rest through 20 feet under an acceleration denoted by 32, the equation—
$$v^2 = 2gs$$
becomes in this case—
$$v^2 = 64 \times 20$$
$$= 1280.$$

EXAMPLES ON COLLISION. 203

If the *upward* velocity of the ball after impact be v' feet per second, we have $v' = ev$, and therefore—

$$v'^2 = \frac{3^2}{4^2} \cdot 1280$$

$$= 720.$$

Also, if h be the height to which it rises in the rebound,

$$h = \frac{v'^2}{2g},$$

or $\quad h = \dfrac{720}{64} = 11\frac{1}{4},$

or the ball rises to a height of $11\frac{1}{4}$ feet.

222. Ex. 3. *A particle is projected from a point distant 20 feet from a rough vertical wall, with a velocity of 60 feet per second, and at an elevation of 60°, its plane of motion being perpendicular to that of the wall. Find its velocity and the direction of its motion immediately after striking the wall, the coefficient of elasticity being $\dfrac{3}{5}$, and the coefficient of friction $\dfrac{1}{2}$.*

The initial velocity of the particle may be resolved into a velocity of 30 feet per second in a horizontal direction, and $60\,\dfrac{\sqrt{3}}{2}$ or $30\sqrt{3}$ feet per second vertically upwards. Now its horizontal velocity remains constant until it strikes the wall. Hence the time elapsing before it strikes the wall, that is, while it moves over a horizontal distance of 20 feet, is two-thirds of a second. Its vertical velocity when it strikes the wall is $30\sqrt{3} - \dfrac{2}{3}g$ feet per second.

Now, since the coefficient of elasticity is $\dfrac{3}{5}$, its horizontal velocity after impact is 18 feet per second away from the wall. Hence the whole change of its horizontal velocity is represented by 48 feet per second, and if m denote the mass of the particle, the impulse of the pressure on the wall will be represented by $m.48$. Now, when the particle strikes the wall, the component of its velocity parallel to the wall is $30\sqrt{3} - \frac{2}{3}g$ feet per second vertically upwards. Hence the impulsive friction acts vertically downwards; and since the coefficient of friction is $\frac{1}{2}$, the impulsive friction is one-half the impulsive pressure, and its impulse is therefore numerically equal to $24m$, and this corresponds to a downward velocity of 24 feet per second. The vertical velocity of the particle after impact will therefore be a velocity of $30\sqrt{3} - \frac{2}{3}g - 24$ feet per second in the upward direction. Hence, after impact, the vertical component of the particle's velocity is 6·63 feet per second very nearly, and the horizontal component is 18 feet per second. If θ be the angle which its direction of motion makes with the horizon, we have—

$$\tan \theta = \frac{221}{600} \text{ very nearly,}$$

and the velocity of the particle is nearly $\sqrt{18^2 + 6\cdot3^2}$ feet per second.

If the friction found as above had been more than sufficient to destroy the particle's upward velocity, only sufficient friction would have been called into play to destroy this, and the particle after impact would move in a direction perpendicular to the wall, unless the friction possess a property of the nature of elasticity.

223. **Ex. 4.** *A straight staircase contains any number of stairs, each one foot wide and six inches high. A small smooth ball is projected from a point on one of the stairs near its edge, and in the vertical plane perpendicular to the edge of each stair. Find the velocity and elevation of projection in order that it may strike each succeeding stair at the same distance from the edge, the coefficient of elasticity being* $\frac{1}{2}$.

EXAMPLES ON COLLISION. 205

The given condition is of course that which must be fulfilled in order that the particle may bound down an unlimited number of stairs, striking each only once.

Let A be the point of projection, B, C_1... the points where the particle strikes the succeeding stairs; let θ be

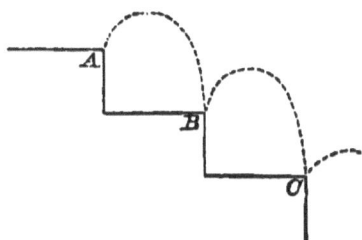

the elevation of projection, and v the velocity in feet per second. Then the horizontal velocity of the particle is constant and equal to $v \cos \theta$. Its vertical velocity is initially $v \sin \theta$. Now since the width of each stair is one foot, and the horizontal velocity constant and equal to $v \cos \theta$ feet per second, the time occupied by the particle in passing from A to B is $\dfrac{1}{v \cos \theta}$ seconds. But the vertical space through which the particle rises in t seconds is $v \sin \theta \cdot t - \dfrac{1}{2} g t^2$. Hence, in order that the particle may reach B, we must have—

$$v \sin \theta \cdot \frac{1}{v \cos \theta} - \frac{1}{2} g \frac{1}{v^2 \cos^2 \theta} = -\frac{1}{2},$$

since each stair is 6 inches high;

$$\therefore v^2 \sin 2\theta + v^2 \cos^2 \theta = g \quad \ldots\ldots\ldots\ldots\ldots\ldots(\text{I.})$$

This is the first condition required.

Now, since the ball is smooth, its horizontal velocity, after striking B, is the same as before; viz., $v \cos \theta$. Hence if its vertical velocity when leaving B be the same as when projected from A, the direction of the ball's motion on leaving B will be the same as when projected from A, and its whole velocity will be the same. Hence the ball will strike the next stair as required, and everything being the same as before, it will leave C with the same velocity and in the same direction as it left B, and so on for any number of stairs. We see, then, that the only other condition which we have to satisfy, besides that given in (I.), is, that the vertical component of the ball's velocity on leaving B should be the same as on leaving A.

The vertical velocity of the ball downwards on reaching B is $gt - v \sin \theta$, if t be the time of flight from A to B. But t is equal to $-\dfrac{g}{v \cos \theta}$. Hence the vertical component of the ball's velocity when it reaches B is $\dfrac{g}{v \cos \theta} - v \sin \theta$. But since the coefficient of elasticity is $\dfrac{1}{2}$, its vertical velocity on leaving B is numerically equal to one-half of this. Hence we must have—

$$\dfrac{1}{2}\left(\dfrac{g}{v \cos \theta} - v \sin \theta\right) = v \sin \theta,$$

or
$$\dfrac{g}{v \cos \theta} = 3v \sin \theta;$$

$$\therefore v^2 \sin 2\theta = \dfrac{2}{3}g \dots\dots\dots\dots\dots\dots\text{(II.)}.$$

This equation, together with equation (I.), determines v and θ. From (I.) and (II.) we have—

$$v^2 \cos^2 \theta = \dfrac{1}{3}g;$$

$$\therefore \sin 2\theta = 2 \cos^2 \theta;$$
$$\therefore \sin \theta = \cos \theta;$$
$$\therefore \theta = 45°$$

Hence $\sin 2\theta = 1$,

and we have $v^2 = \dfrac{2}{3}g;$

$$\therefore v = \sqrt{\dfrac{2}{3}g}.$$

EXAMPLES ON COLLISION. 207

If we take g equal to 32, we get—

$$v = \frac{8}{\sqrt{3}},$$

or the velocity of projection must be $\frac{8}{\sqrt{3}}$ feet per second, and the elevation of projection we have seen to be 45°.

224. **Ex. 5.** *An engine whose mass is 40 tons, and 20 coal-trucks, each of 15 tons, are at rest on a horizontal line, there being an interval of one foot between the engine and the first truck, and between each truck and the next succeeding. The engine starts off and strikes the first truck, which then strikes the second, and so on down the train, the trucks being each inelastic. Supposing the engine to be constantly impelled by a force equal to the weight of one ton, find the velocity with which the last truck starts and the whole time occupied in starting the train, neglecting friction, and taking g equal to 32.*

Let v' be the velocity of the engine when it strikes the first truck. Then, since the mass of the engine is 40 tons, and it is acted upon by a force equal to the weight of 1 ton, it moves over the one foot between its initial position and the first truck with a constant acceleration denoted by $\frac{g}{40}$. Hence we have by Art. 152—

$$v'^2 = 2\frac{g}{40}.$$

Now, when it strikes the first truck, the two will proceed with a common velocity. Let v_1 denote this velocity. Then, since the momentum of the engine and truck immediately after impact is equal to that of the engine before impact, we have—

$$(40 + 15)v_1 = 40v',$$

or

$$v_1^2 = \left(\frac{40}{55}\right)^2 \cdot \frac{2g}{40}$$

$$= \frac{2g \cdot 40}{55^2}.$$

We have now a mass of 55 tons impelled by a constant force equal to the weight of one ton, and therefore moving with constant acceleration $\frac{g}{55}$. Hence if v_1' denote its velocity when it strikes the second truck, since the space over which it has passed under this acceleration is one foot, we have—

$$v_1'^2 - v_1^2 = \frac{2 \cdot g}{55}.$$

Therefore $\quad v_1'^2 = \dfrac{2g \cdot 40 + 2g \cdot 55}{55^2}.$

Let v_2 denote the common velocity immediately after striking the second truck; then, since the whole momentum is unaltered,—

$$(55 + 15)\, v_2 = 55 v_1'.$$

Therefore

$$v_2^2 = \left(\frac{55}{70}\right)^2 v_1'^2$$

$$= \left(\frac{55}{70}\right)^2 \cdot \frac{2g \cdot (40 + 55)}{55^2}$$

$$= 2g \cdot \frac{40 + 55}{70^2}.$$

We have now a mass of 70 tons moving under a constant force equal to the weight of one ton. Hence if v_2' denote its velocity when the third truck is struck,—

$$v_2'^2 - v_2^2 = \frac{2g}{70};$$

$$\therefore v_2'^2 = 2g \cdot \frac{40 + 55 + 70}{70^2}.$$

When the third truck is struck, the mass in motion is changed from 70 tons to 85 tons. Hence if v_3 denote the common velocity immediately after striking,

$$v_3 = \frac{70}{85} v_2'.$$

EXAMPLES ON COLLISION.

or
$$v_3{}^2 = \left(\frac{70}{85}\right)^2 . 2g . \frac{40+55+70}{70^2}$$
$$= 2g . \frac{40+55+70}{85^2}.$$

Proceeding in this way, we see that the common velocity, v_{19}, immediately after starting the 19th truck, is given by the equation—

$$v_{19}{}^2 = 2g . \frac{40+55+70+\ldots+310}{325^2},$$

and if v_{19}' be the velocity of the rest of the train at the instant when the last truck is struck,

$$v_{19}'^2 - v_{19}{}^2 = \frac{2g}{325},$$

or
$$v_{19}'^2 = 2g . \frac{40+55+70+\ldots+310+325}{325^2}.$$

But if v_{20} denote the common velocity immediately after starting the last truck,—

$$340 v_{20} = 325 v_{19}',$$

or
$$v_{20}{}^2 = \left(\frac{325}{340}\right)^2 . 2g . \frac{40+55+70+\ldots+325}{325^2}$$
$$= 2g . \frac{40+55+70+\ldots+325}{340^2}$$
$$= 2g . \frac{3650}{340^2}$$
$$= 64 . \frac{3650}{340^2};$$
$$\therefore v_{20} = \frac{8 . \sqrt{3650}}{340}$$
$$= 1\cdot 421\ldots$$

Hence the last truck starts with a velocity of $1\cdot 421$ feet per second, very nearly.

G. D.

Since the force producing motion is always equal to the weight of a ton, if we take a ton for the unit of mass, the momentum generated in t seconds will be represented by gt. Now *since the momentum of the train is unaltered by the successive impacts*, the momentum of the train when the last truck starts must be that produced by a force equal to the weight of a ton in the time t, if t be the time which has elapsed since the engine started. Hence since the mass of the train is 340 tons,

$$340 v_{20} = gt,$$

$$\text{or} \quad t = \frac{340 v_{20}}{g}$$

$$= \frac{8\sqrt{3650}}{32}$$

$$= 15\cdot 103\ldots,$$

or the time required to start the train is rather more than 15·1 seconds.

225. Suppose n equal particles, each of mass m, and moving with a velocity v, to impinge directly upon a fixed inelastic plane surface during t seconds. The surface will in the course of the t seconds receive n impulses, each represented by mv, and the whole momentum destroyed by the surface in the t seconds is represented by nmv. Now suppose the intervals between successive impacts to be equal to one another, and the velocity of each particle remaining the same, suppose n to increase while m diminishes in the inverse ratio, so that mn remains constant and equal to M say. Then the sum of the impulses upon the surface during t seconds is the same as before, and would generate a momentum represented by Mv, and this is true, however great n may be. But if n become indefinitely great, we can no longer distinguish any interval between successive impulsive pressures; in fact the action upon the surface becomes a continuous and uniform pressure, and the momentum destroyed in t seconds by the reaction of the surface, which is always equal to this pressure, is Mv. Hence the momentum which would be destroyed in one second by this pressure is equal to

CONTINUOUS IMPACT. 211

$\frac{Mv}{t}$, which is therefore the measure of the pressure. The reaction of the surface is, in fact, measured at any time by the rate at which momentum is being destroyed by it.

226. The pressure of the wind, or of a jet of water upon any object which it strikes, may be taken as an illustration of continuous pressure produced by a quick succession of impacts. The air, or water, which comes in contact with the obstacle in the course of a second consists of a very great number of particles, each of which strikes the obstacle with a certain velocity, and the impacts are in such quick succession that they link themselves together into a continuous pressure.

227. In the case of a jet of water striking a wall, if the area of the section of the jet remains always the same, the amount of matter which strikes the wall in a second will be proportional to the velocity; and since the momentum of each particle of water is proportional to its velocity, it follows that the change of momentum produced by the reaction of the wall in each second is proportional to the square of the velocity of the water, supposing this constant, and hence the pressure on the wall is proportional to the square of the velocity of the jet.

228. For example, suppose a jet of water, the area of whose transverse section is one square inch, to impinge directly upon a wall with a velocity of 128 feet per second, the coefficient of elasticity being $\frac{1}{16}$. We proceed to find the pressure on the wall.

Since the area of the section of the jet is one square inch and the velocity of the water 128 feet per second, the volume of water which strikes the wall in one second is 128 × 12 cubic inches, and since one cubic foot of water contains 1000 ozs., the mass of this is $\frac{128 \times 12}{1728}$. 1000 ozs. or $\frac{500}{9}$ pounds. Now the velocity of this before impact is 128 feet per second towards the wall, and after impact it is $\frac{128}{16}$ feet per second away from the wall, since the coefficient of elasticity is $\frac{1}{16}$. Hence the change of velocity produced by the impact is 136 feet per second. The change of momentum produced by the reaction of the wall in one second is therefore $\frac{136 \times 500}{9}$ units of momentum. But the weight of one pound generates in the mass of one pound in one second a velocity of 32 feet per second, or the weight of one pound generates in one second 32 units of momentum. The reaction of the wall is therefore equivalent to the weight of $\frac{136 \times 500}{9 \times 32}$ pounds, and since action and reaction are equal and opposite, it follows that the pressure of the jet upon the wall is equal to the weight of $\frac{136 \times 500}{9 \times 32}$ pounds, that is, of $236\frac{1}{9}$ lbs.

This result affords an explanation of the great mechanical effect produced by the jet from a fire-engine. The result that the pressure varies as the square of the velocity is the basis of the ordinary theory of fluid resistances.

229. *A perfectly flexible uniform chain, the mass of each unit of length of which is* m, *hangs vertically from its upper extremity with the lower end just in contact with an inelastic table. If the chain be allowed to fall, find the pressure upon the table at any instant during the motion.*

If a series of particles are arranged in a vertical line, and then all allowed to fall freely at the same instant, their velocities at any subsequent time will be the same, and the distance through which they will have fallen will be the same for each. Hence they will always remain in a vertical straight line at the same distance apart as when they started, although they are perfectly free. Consequently if they be so connected together that they are unable to alter their distances from each other, there will be no stress on the connections and the motion will be unaffected thereby. Hence, in the case of the falling chain, each particle of the chain will fall as if it were free. Therefore at the end of t seconds, after the motion has commenced, the velocity of each particle of the chain will be gt, and the space through which the upper part will have fallen will be $\frac{1}{2}gt^2$. There will therefore be a length of chain measured by $\frac{1}{2}gt^2$ coiled up upon the table. Now if the velocity of the chain were to remain constant during one second, and the same as at the time t, the length of chain which would be brought to rest upon the table during that second would be gt, and its mass mgt. Also, its velocity being gt, the momentum which would be destroyed during the second would be mg^2t^2. Hence mg^2t^2 is the expression for the rate at which momentum is being destroyed at the end of t seconds after the commencement of the motion. But the rate of change of momentum is the measure of the force producing that change. Hence at the end of the time t the reaction of the table required to destroy the momentum of the falling chain is numerically equal to mg^2t^2, and therefore the pressure exerted by the falling portion of the chain in coming to rest upon the table is denoted by mg^2t^2. But at the end of the time t there is a length of chain, denoted by $\frac{1}{2}gt^2$, coiled up on the table, and the weight of this is

$$mg \cdot \frac{1}{2}gt^2, \quad = \frac{1}{2}mg^2t^2.$$

Hence the whole pressure upon the table is $\frac{3}{2}m.g^2t^2$, that is, three times the weight of the portion of chain coiled up.

If the lower end of the chain had been at a height h above the table, the length of chain coiled up at the end of t seconds would have been $\frac{1}{2}gt^2 - h$, provided $\frac{1}{2}gt^2$ were greater than h, and the whole pressure upon the table would have been

$$\frac{3}{2}mg^2t^2 - mgh.$$

230. *A perfectly flexible uniform chain, the mass of each unit of length of which is* m, *is coiled up in the hand, and one end is attached to a fixed point. Suddenly the hand is removed; it is required to find the force upon the fixed point at any instant before the whole of the chain has come to rest.*

Each particle of the chain which remains in the coil will, at any subsequent time, be falling freely, but successive portions of the coil will be brought to rest, and hang vertically from the fixed point. The velocity of every particle in the coil at the end of the time t after the commencement of the motion will be gt, and the space through which the coil will have fallen will be denoted by $\frac{1}{2}gt^2$. Hence the weight of chain hanging from the fixed point will be $\frac{1}{2}mg^2t^2$. Also, as in the preceding example, the rate of destruction of momentum will be mg^2t^2. Hence the whole force upon the fixed point will be $\frac{3}{2}mg^2t^2$, or three times the weight of chain hanging vertically from the point.

The tension at any point of the chain distant h below the highest point will be less than the tension at the highest point by the weight of the length h of the chain, and will therefore be denoted by $\frac{3}{2}mg^2t^2 - mgh$.

231. We will give two other examples of falling chains.

A flexible chain is suspended from a fixed point, and hangs vertically with its lower end just touching an inelastic horizontal table; it is then allowed to fall. Supposing the density at any point of the chain to be proportional to the distance from the lower end, find the pressure on the table at any subsequent time.

As in the preceding example, the chain will fall freely, and its velocity at the end of t seconds from the commencement of the motion will be gt, while the length of chain coiled upon the table will be $\frac{1}{2}gt^2$.

FALLING CHAIN OF VARIABLE DENSITY. 215

Let the density of the chain at a point distant s from the lower end be $m\dfrac{s}{h}$. This, then, would be the mass of a unit of length of the chain, supposing its density uniform and the same as at the distance s from the lower end. Also the mass of a thin section of the chain of thickness k at this point is $m\dfrac{s}{h}.k$, since we may suppose the density uniform throughout the section, k being very small. Now if b be the length of the base of a triangle and h its altitude, the area of an indefinitely narrow strip of breadth k at distance s from the vertex is $b\dfrac{s}{h}.k$. Hence the problem of finding the mass of any length s of the chain is the same as that of finding the area of the triangle cut off from the above-mentioned triangle by a line parallel to the base, and distant s from the vertex, and the area of this triangle is $\dfrac{1}{2}b\dfrac{s}{h}.s$, or $\dfrac{1}{2}\dfrac{b}{h}s^2$. Therefore the mass of a length s of the chain measured from the lower end is $\dfrac{1}{2}m\dfrac{s^2}{h}$. The weight of the chain upon the table at the end of the time t is therefore

$$\frac{1}{2}mg \cdot \frac{1}{h}\left(\frac{1}{2}gt^2\right)^2, \text{ or } \frac{1}{8}m\frac{g^3t^4}{h}.$$

The density of the chain at the point which is just coming to rest is $m\dfrac{1}{2}\dfrac{gt^2}{h}$, and its velocity is gt. Now if the velocity of the chain remained uniform throughout one second and equal to gt, and if the density of the chain which comes to rest during that second were uniform, and the same as at the point which is just coming to rest at the end of the time t, the mass which would be reduced to rest during the second would be the mass of a length gt of the chain, the mass of each unit of length of which would be $\dfrac{1}{2}m\dfrac{gt^2}{h}$, that is, $\dfrac{1}{2}m\dfrac{g^2t^3}{h}$ units of mass. Also the velocity of this being gt the momentum which would be destroyed during the second would be $\frac{1}{2} m \frac{g^3 t^4}{h}$. Hence the rate at which momentum is being destroyed at the end of the time [t is $\frac{1}{2} m \frac{g^3 t^4}{h}$ units of momentum destroyed per second, and the force required to do this is measured by $\frac{1}{2} m \frac{g^3 t^4}{h}$. But the weight of chain upon the table is $\frac{1}{8} m \frac{g^3 t^4}{h}$. Hence the whole pressure upon the table is $\frac{5}{8} m \frac{g^3 t^4}{h}$, that is, five times the weight of the chain coiled upon the table.

232. *Suppose the chain described in the last Article to be coiled up close to the edge of a smooth table, the end, which in the last example was in contact with the table, being allowed to hang just over the edge. If the chain is then allowed to run off the edge, find the motion, and the tension close to the edge of the table, at any subsequent time.*

Suppose at the end of the time t there is a length s of the chain in motion, and let v be its velocity. Then the chain is being pulled off the table at the *rate* of v units of length, or of $m \frac{s}{h} v$ units of mass, per second, and the velocity with which each particle is being started is v. Hence the rate at which momentum is being generated in the chain just coming off the table is measured by $m \frac{s}{h} v^2$ units of momentum per second. The tension of the chain close to the edge of the table is therefore represented by $m \frac{s}{h} v^2$, and this tension is acting upwards upon the length s of chain which is hanging over the edge and descending vertically. Now the mass of this length s of the chain is $\frac{1}{2} m \frac{s^2}{h}$, and its weight is therefore $\frac{1}{2} mg \frac{s^2}{h}$. Hence the resultant force upon this portion of the chain is a downward force represented by $\frac{1}{2} mg \frac{s^2}{h} - m \frac{s}{h} v^2$, and the acceleration produced by this force in the mass $\frac{1}{2}m\frac{s^2}{h}$ will be represented by $g - \frac{2v^2}{s}$. Therefore, when a length s of the string is in motion, if v be its velocity, the acceleration under which it is moving is $g - \frac{2v^2}{s}$. But the velocity as well as the space passed over in any time depends upon the acceleration during each instant of the motion. If then we can determine an expression for the acceleration so that throughout the motion $g - \frac{2v^2}{s}$ may be numerically equal to it, the acceleration so determined will be that under which the chain is moving. Now if a particle be moving under constant acceleration f, the velocity v at any time t will be equal to ft, and the space s passed over will be represented by $\frac{1}{2}ft^2$. Hence in this case $\frac{v^2}{s} = 2f$ and is constant. Hence if the end of the chain were moving with a constant acceleration f we should have $g - \frac{2v^2}{s}$ equal to $g - 4f$, that is, a constant, whatever be the time which has elapsed since the beginning of the motion, and we can obviously determine f so that this expression may be equal to f. But this is the only condition which the motion of the chain has to fulfil. Hence if f be so determined, it will be the acceleration under which the chain is descending. To determine it we have

$$g - 4f = f;$$
$$\therefore f = \frac{1}{5}g,$$

or the end of the chain descends with uniform acceleration equal to one-fifth of that produced by gravity in a free particle.

Hence the velocity v at the end of t seconds after the commencement of motion is given by

$$v = \frac{1}{5}gt,$$

and the space s through which the end of the chain has descended is given by

$$s = \frac{1}{10} g t^2$$

The weight of the portion of chain in motion is $mg\frac{s^2}{2h}$, that is

$$m \cdot \frac{g^3 t^4}{200 h},$$

and the tension at the point where it leaves the table is equal to $m\frac{s}{h}v^2$, that is, to

$$m \cdot \frac{g^3 t^4}{250 h},$$

or the tension at the point where the chain leaves the table is four-fifths of the weight of the portion in motion. This agrees with the result that the acceleration of the falling portion is $\frac{1}{5}g$, that is, that the resultant force upon it is one-fifth of its weight.

233. *Two balls, whose masses are* M *and* m *respectively, moving in the same straight line, impinge directly upon one another. It is required to find the change of kinetic energy produced by the impact.*

Let V, v be their respective velocities before impact, V_1, v_1, their velocities after impact. Then, by Article (214), we have

$$V_1 = \frac{MV + mv - em(V-v)}{M+m},$$

$$v_1 = \frac{MV + mv + eM(V-v)}{M+m}.$$

Let E denote the kinetic energy of the two before impact, E_1 their energy after impact. Then

$$E = \frac{1}{2}(MV^2 + mv^2),$$

KINETIC ENERGY AFTER IMPACT.

and

$$E_1 = \frac{1}{2}(MV_1^2 + mv_1^2)$$

$$= \frac{M}{2}\left\{\frac{MV + mv - em(V-v)}{M+m}\right\}^2 + \frac{m}{2}\left\{\frac{MV + mv + eM(V-v)}{M+m}\right\}^2$$

$$= \frac{(M+m)(MV+mv)^2 + (m+M)e^2 Mm(V-v)^2}{2(M+m)^2}$$

$$= \frac{(MV+mv)^2 + e^2 Mm(V-v)^2}{2(M+m)}.$$

If e be equal to 1, or the elasticity perfect,

$$E_1 = \frac{(MV+mv)^2 + Mm(V-v)^2}{2(M+m)}$$

$$= \frac{M^2V^2 + m^2v^2 + MmV^2 + Mmv^2}{2(M+m)}$$

$$= \frac{MV^2 + mv^2}{2}$$

$$= E.$$

Hence, if the elasticity be perfect, the total kinetic energy of the two balls is the same after impact as before.

If e be less than 1, we may put the expression for E_1 into the form—

$$E_1 = \frac{(MV+mv)^2 + Mm(V-v)^2}{2(M+m)} - \frac{(1-e^2)Mm(V-v)^2}{2(M+m)}.$$

But the first term on the right-hand side we have seen to be equal to E; hence

$$E_1 = E - \frac{(1-e^2)Mm(V-v)^2}{2(M+m)};$$

and the second term on the right-hand side of this last

equation is necessarily positive. Hence in the direct impact of imperfectly elastic balls an amount of kinetic energy denoted by $(1-e^2)\dfrac{Mm}{2(M+m)}(V-v)^2$ is lost.

This kinetic energy is chiefly transformed into molecular vibrations in the balls, and becomes sensible to us in the form of heat.

234. If the balls considered in the preceding article impinge obliquely upon one another, then, resolving their velocities as in Art. (216) along and perpendicular to the line joining their centres, if a, β be the angles which their directions of motion make with this line before impact, and θ, ϕ similar quantities after impact, we have from Art. (216)

$$\left. \begin{array}{l} V_1 \sin \theta = V \sin a \\ v_1 \sin \phi = v \sin \beta \end{array} \right\} \quad \ldots\ldots\ldots\ldots\ldots\ldots(I.)$$

Also

$$\left. \begin{array}{l} V_1 \cos \theta = \dfrac{MV\cos a + mv \cos \beta - em(V \cos a - v \cos \beta)}{M+m} \\ v_1 \cos \phi = \dfrac{MV\cos a + mv \cos \beta + eM(V \cos a - v \cos \beta)}{M+m} \end{array} \right\} \ldots(II.)$$

Now the kinetic energy of the two balls before impact is

$$\tfrac{1}{2}(MV^2 + mv^2), \text{ that is}$$

$$\tfrac{1}{2}\{MV^2(\cos^2 a + \sin^2 a) + mv^2(\cos^2 \beta + \sin^2 \beta)\}.$$

Let us denote this quantity by E. Then if E_1 denote the kinetic energy after the impact

$$E_1 = \tfrac{1}{2}\{MV_1^2(\cos^2\theta + \sin^2\theta) + mv_1^2(\cos^2\phi + \sin^2\phi)\}.$$

Now from equations (I.)

$$\tfrac{1}{2}(MV_1^2 \sin^2\theta + mv_1^2 \sin^2\phi) = \tfrac{1}{2}(MV^2 \sin^2 a + mv^2 \sin^2 \beta) \ldots(III.)$$

and, precisely as in the preceding article, we may show from equations (II.) that

$$\frac{1}{2}(MV_1^2\cos^2\theta + mv_1^2\cos^2\phi)$$
$$= \frac{1}{2}(MV^2\cos^2\alpha + mv^2\cos^2\beta) - \frac{(1-e^2)Mm(V\cos\alpha - v\cos\beta)^2}{2(M+m)} \quad \text{(IV.)}$$

since we have only to write $V\cos\alpha$, $v\cos\beta$, $V_1\cos\theta$, $v_1\cos\phi$ for V, v, V_1 and v_1 respectively in the equations of that article.

Hence, adding equations (III.) and (IV.), we get

$$E_1 = E - (1-e^2)\frac{Mm}{2(M+m)}(V\cos\alpha - v\cos\beta)^2.$$

As before, we see that if the elasticity be perfect the kinetic energy is the same after impact as before, but if it be imperfect there is a loss of energy by the impact.

235. Suppose we have a cylinder of some compressible material, say one foot in length, and suppose that when a pressure equal to the weight of 100 lbs. is applied at the ends the length is diminished by ·01 inch. Then, provided we keep within the limits of elasticity, a pressure equal to the weight of 200 lbs. would shorten the cylinder by nearly ·02 inch, and so on in proportion. The weight of one pound would diminish the length of the cylinder by ·0001 inch, or by $\frac{1}{120000}$ of the original length. If the cross section of the cylinder be equal to the unit of area the elasticity of the material as measured by the quotient $\frac{\text{stress}}{\text{strain}}$ becomes 120000, the weight of a pound being taken as unit of force.

236. Now suppose that the cylinder has been shortened by the application of pressure at the ends, and let the pressure now be gradually diminished. Suppose that as soon as the pressure is diminished ever so little the length of the cylinder begins to increase, and increases in such a way, as the pressure continues to be diminished, that the length under any given pressure is exactly the same as

under the same pressure when the cylinder was being compressed, and that this continues until the cylinder has regained its original length. In this case just as much work will be done by the elastic force during the restitution of form of the cylinder as was done upon the cylinder during the compression. Such a cylinder might be alternately compressed and allowed to expand without any loss of mechanical energy, and without any heat being produced. Its elasticity or power of regaining its original form may be considered as perfect.

237. On the other hand, it may happen that when the body has been compressed, though a very great pressure may have been required to produce the deformation, yet the whole of this pressure may be removed without the body showing any tendency to return to its original form. In such a body the elasticity, or power of restitution of form, may be considered as zero, and nearly the whole of the work done upon the body during its compression has its equivalent generally in heat produced within the body.

238. Between these two extremes we may have any number of intermediate links. Suppose, for example, that in the case assumed above the cylinder has been loaded with 1000 lbs. and therefore shortened by ·1 inch. Now suppose that the pressure is diminished, and that the cylinder retains its length of 11·9 inches until the pressure has been reduced to 500 lbs.' weight, and that it only begins to return when the pressure is reduced below this. Suppose also that throughout the restoration of form the pressure corresponding to any length of the cylinder is exactly half what it was during the compression, so that when the length of the cylinder is 11·95 in. the pressure is equal to the weight of 250 lbs., and when the length is 11·99 in. the pressure is equal to the weight of 50 lbs., and so on. Then exactly one-half of the amount of work done in compressing the cylinder will be restored by the elastic forces during the restoration of form, the other half being mostly converted into heat within the imperfectly elastic material. Generally, if the pressure at any stage during the restoration of form be n times that at the correspond-

KINETIC ENERGY AFTER IMPACT. 223

ing stage of the compression, the work done by the elastic forces during the restoration will be n times that done against them during the compression, while the energy permanently transformed into heat or otherwise wasted (as far as mechanics is concerned) will be $1-n$ times the work done in compressing the body.

239. Now suppose that two bodies are compressed by impinging against each other, and suppose, for the sake of simplicity, that they are both of the same material, whose elasticity is such that the pressure exerted at any stage during the restoration of form is n times that required to produce further compression at the same stage, and that the restoration of form is complete. Now let E denote the number of units of work done in compressing the bodies up to the condition of maximum compression, that is, up to the instant when the relative motion of the centres of gravity of the bodies is zero. Then up to this instant E units of kinetic energy will have been lost by the moving bodies, having been expended in doing work in producing compression. During the restoration of form nE units of work will be done by the elastic forces and expended in producing motion in the bodies, so that nE units of kinetic energy will be returned to the system; the remaining $(1-n)E$ units of energy being converted into heat or otherwise disposed of. Comparing these expressions with those obtained in Art. 233, for the kinetic energy lost by impact, we see that n must be equal to e^2. Hence e, the coefficient of elasticity, must be equal to the square root of the ratio of the pressures between the balls at the same stage of the restitution and compression respectively. Hence e is equal to the *square root* of the ratio of the true force of restitution to the true force of compression in any, the same, configuration. To use the terms "force of restitution" and "force of compression" in the sense explained in Art. 217 is consequently not only to use words in their wrong senses, but to convey a false impression of what the relation between the forces of restitution and compression really is.

240. The coefficient of elasticity, e, is the ratio between the *impulses* of the forces of restitution and com-

224 ENERGY DISSIPATED BY IMPULSIVE ACTIONS.

pression, and differs from the ratio of the forces, since the time occupied by the compression is only e times the time occupied by the restitution. The basis of this last statement we will now consider. From the instant of greatest compression up to that of complete restoration of form the centres of gravity of the bodies move relatively to one another over exactly the same distance as from the commencement to the end of the compression. Also at any stage in the restoration the pressure between the balls is n times as great as at the same stage (*i.e.*, the same configuration) during the compression. Hence the acceleration is n times as great as the retardation during the compression. Now the time taken to travel over any distance from rest with uniform acceleration is inversely proportional to the square root of the acceleration (for $s = \frac{1}{2} ft^2$), and it may be easily shown geometrically that if two points travel over the same distance from rest with accelerations which vary in any manner, but which bear a constant ratio, n, to one another, when the points are in the same positions, then the times taken by two points to travel the same distance will be to one another in the ratio of 1 to $\sqrt{n}$. Also the velocities at the end of the times will be in the ratio of $\sqrt{n}$ to 1, for though the first acceleration is n times that of the second, the time during which the velocity of the first point is being increased is $\frac{1}{\sqrt{n}}$ times that during which the velocity of the second is being increased, so that the final velocities generated are in the ratio of $\sqrt{n} : 1$. If the points be replaced by material particles, the forces upon them will be in the ratio of $n : 1$. The quantities of work done by these forces, since the distance travelled is the same for both, will be in the ratio of $n : 1$, the velocities generated will be in the ratio of $\sqrt{n} : 1$; and therefore the quantities of kinetic energy generated will also be in the ratio of $n : 1$, that is, in the same ratio as the quantities of work done by the forces, which is in accordance with the principle of the conservation of energy.

241. It is important to notice that whenever a portion

of matter, however small it may be, has its velocity *suddenly* changed by any means whatever, there is, in all cases, a loss (or transformation) of energy, unless all the portions of matter concerned in the action be *perfectly elastic*. The energy thus "lost" is generally converted into heat, and whenever actions of this kind take place we cannot employ the method of conservation of energy unless we take account of the energy thus transformed. In such cases it is generally best to solve the problem from the ordinary equations of motion.

242. A familiar illustration is afforded by heavy flexible strings or chains, some examples of which we have already considered. Thus if a string be suspended above an inelastic table, and then allowed to fall, as the successive elements of the string strike the table and are brought to rest, the table being supposed inelastic, the energy of their motion is converted into heat, and this is similar to other cases of inelastic collision. But if, when the string is lying in a coil on the table, we take hold of one end, and raise it with uniform velocity, there is also a transformation of energy into heat, so that the work we have to do is greater than that done against gravity in raising the string together with the kinetic energy of the moving portion of the string. Thus, if we consider a particle of the string at rest upon the table, it remains at rest until its turn comes to move, and then it starts off with a jerk, the action between it and the particle next above it being similar to the collision of inelastic bodies, except that it is an impulsive *tension* instead of *pressure* which takes place, and energy is dissipated, heat being produced by the action.

243. The following example will further illustrate this point.

EXAMPLE. *A coil of heavy rope of indefinite length lies on the ground while one end of the rope passes round a windlass, and is coiled on a scaffold at a height* h *above the ground. The windlass is turned by an engine which raises the rope with uniform velocity* v *from the ground to the scaffold. If*

226 ENERGY DISSIPATED BY VARIATION OF STRESS.

m *be the mass of unit of length of the rope, find the rate at which the engine works.*

[We might be disposed to treat this question as follows:—

In one second a weight mgv of string is raised through a height h, and therefore the work done *against gravity* is $mgvh$.

But in each second mv units of mass of the string are started with velocity v, and therefore receive a quantity of energy represented by $\frac{1}{2}mv \cdot v^2$ or $\frac{1}{2}mv^3$, which is converted into heat as the motion of the rope is destroyed on reaching the upper coil. Hence the whole amount of work done in a second is

$$mgvh + \frac{1}{2}mv^3.$$

In this investigation we have omitted the energy which is transformed into heat at the point where the rope leaves the first coil, and where the particles of the rope are made to move off with a jerk. The following is the safe mode of procedure.]

The weight of rope hanging from the windlass on one side in excess of that hanging on the other side is mgh.

In each second mv units of mass of rope are made to move from rest with velocity v, and therefore acquire mv^2 units of momentum. To produce this, mv^2 units of force must be exerted.

Hence the tension of the rope on one side of the windlass must exceed that on the other side by $mgh + mv^2$ units of force; and since the rope is raised with velocity v the engine overcomes a resistance of $mgh + mv^2$ units of force through v units of length in one second and the rate of doing work is

$$mghv + mv^3.$$

From this it will be seen that just as much energy is transformed into heat in starting the rope from the lower coil as in bringing it to rest in the upper coil, the amount in each case being $\frac{1}{2}mv^3$ units per second.

244. When a stress is applied to a body and the material yields, as all materials must yield and suffer deformation or strain, a certain amount of work is done in straining the body against its elastic forces. When the stress is relieved and the body returns to its original form, as it will do if the stress applied has not exceeded its elastic limits, more or less, but never quite the whole, of the energy absorbed in straining the body is restored. In the case of highly tempered steel, such as tuning forks are made of, there is very little energy dissipated when the steel is strained and then allowed to return to its original form. In other cases the amount of energy wasted is much greater. In most machinery the parts are so designed that they may suffer very little strain under the stresses to which they are exposed, and those parts which are subject to varying stress are made relatively stronger than those on which the stress is constant. In any case, however, when the stress varies the material must undergo changes of strain, and this must be accompanied by waste of energy. When the forces are of such great intensity, and act for so short a time, as to come within the class of what is generally known as impulsive forces, the material must of necessity undergo considerable variations of strain and a proportionate amount of energy must be wasted. For this reason it is important that all loose fittings, especially in those portions of machines which experience a reciprocating movement, should be avoided. On account of this waste of energy a series of blows extending over a certain interval of time will not produce the same effect as a steady force acting for the same time and equal in intensity to the time average of the impulses, unless the intervals between the blows are too short to allow of any sensible variation in the state of strain of the body struck. In all cases of the transmission of power a uniform driving force working against a uniform resistance is, when possible, the most economical arrangement, for the intermediate portions of the mechanism are then always in the same condition as regards stress and strain.

EXAMINATION ON CHAPTER IV.

1. Describe what takes place when two elastic balls impinge on one another, and define the coefficient of elasticity of two bodies.

2. If a pound fall from a height of 50 feet to the ground, what is the measure of the impulse of the pressure which it will exert, supposing it inelastic?

3. A smooth ball whose mass is 2 oz., moving with a velocity of 10 feet per second, strikes a cushion at an angle of 45°. If the coefficient of elasticity be $\frac{7}{8}$, find the impulse on the cushion and the velocity of the ball after impact.

4. Show that if two perfectly elastic balls of equal mass moving in the same straight line impinge upon one another they will exchange their velocities.

5. A particle strikes a fixed rough plane at an angle of 45°, the coefficient of friction being $\frac{1}{2}$, and the coefficient of elasticity $\frac{1}{2\sqrt{3}}$. The velocity before impact being 20 feet per second, find the velocity and direction of motion after impact.

6. Two smooth billiard-balls moving with equal velocities, v, in directions making an angle of 60° with each other, impinge, the line joining their centres at the moment of impact being at right angles to the direction of motion of one of them. Find the velocity and direction of motion of each after impact, the coefficient of elasticity being $\frac{7}{8}$.

7. In the case of the preceding question find the impulse of the pressure between the two balls.

8. A particle is projected horizontally with a velocity of 40 feet per second from a point 30 feet above a fixed

horizontal plane. Find the height to which it will rise, and its range after the first rebound, the coefficient of elasticity being $\frac{1}{2}$.

9. If a series of perfectly elastic smooth spheres of equal mass be at rest with their centres in a straight line, and be followed by a second series of equal spheres, the mass of each of which is twice that of each of the former, and if the first sphere of the first series be projected so as to impinge directly upon the second, investigate completely the subsequent motion.

10. A stream of water falls from rest at a height of 30 feet above a horizontal inelastic plane at the rate of 100 gallons per minute; find the pressure on the plane, supposing the water to flow freely off it.

11. A ball of 12 pounds, and moving with a velocity of 20 feet per second, impinges directly upon a ball of 20 pounds, moving in the same direction with a velocity of 12 feet per second. Find the amount of kinetic energy lost by the impact, the coefficient of elasticity being zero.

12. An unlimited length of heavy uniform chain is coiled upon a smooth table of height h, one end hanging over the edge and just touching the floor. If the chain be allowed to run down, show that the velocity of the moving portion can never be greater than $\sqrt{gh}$. (See Art. 243.)

EXAMPLES ON CHAPTER IV.

1. An inelastic ball whose mass is 3 lbs. and which moves with a velocity of 30 feet per second, impinges directly on a second inelastic ball of 6 lbs. and moving with a velocity of 8 feet per second in the same straight line. Find their common velocity after impact.

2. In the case of the preceding question find the number of units of work lost by the impact.

3. A shot of 700 pounds, moving with a velocity of

1200 feet per second, enters the side of a ship weighing 6000 tons and remains imbedded in it. Find the velocity which it communicates to the ship.

4. A shot of 700 lbs. is fired with a velocity of 1600 feet per second from a 35 ton gun. Find the velocity with which the gun recoils, neglecting the weight of the powder.

If the recoil of the gun be resisted by a steady pressure equal to the weight of 10 tons, through what space will it recoil?

5. A ball of 40 lbs., moving with a velocity of 80 feet per second, impinges directly on a ball of 100 lbs., moving in the same direction with a velocity of 30 feet per second, their coefficient of elasticity being $\frac{1}{2}$. Find their velocities after impact and the measure of the impulse between them.

6. A ball of elasticity e is dropped from a height h on to a horizontal plane. Show that the whole distance through which it moves before coming to rest is

$$h \frac{1+e^2}{1-e^2}.$$

7. A perfectly elastic billiard-ball impinges on an equal perfectly elastic ball at rest. Show that after impact their directions of motion will be at right angles.

8. One ball impinges on another ball at rest; find the condition that after impact their directions of motion may be at right angles, the coefficient of elasticity being e.

9. A perfectly elastic ball is projected from the middle point of the horizontal base of a vertical square towards one of the upper angles, and after being reflected by both the sides containing that angle, falls at the opposite angle. Determine the velocity of projection.

10. A smooth elastic ball is projected horizontally from the top of a tower 100 feet high with a velocity of 100 feet per second, and after one rebound describes a horizontal range of 40 feet. Find the coefficient of elasticity.

EXAMPLES. 231

11. A particle falls from rest through 16 feet and then rises after impact on a horizontal plane. If the coefficient of elasticity be $\frac{1}{2}$, find its velocity after rising 3 feet and the time of ascending through this height.

12. A perfectly elastic ball falls from a height h on a plane inclined 30° to the horizon; show that it will strike the plane again after an interval equal to twice the time of its fall, and that its range on the plane will be $4h$.

13. If two smooth balls impinge on one another, the motion of their centre of gravity is unaffected by the impact.

14. Two particles whose masses are m, m' are moving in parallel straight lines distant a feet from one another with unequal velocities u and v respectively, and are connected by an inelastic string of length $3a$. Find the impulse of the tension of the string when it becomes tight.

15. Two equal scale-pans, each of mass M, are connected by a string which passes over a smooth peg, and are at rest. A particle of mass m is dropped on one of them from a height $\frac{u^2}{2g}$, the coefficient of elasticity between the scale-pan and particle being e. Find the velocity of the scale-pan after the first impact.

16. Show that, if the string in the preceding example be long enough, the velocity of the scale-pan after the nth impact will be equal to $(1+e)\dfrac{1-e^n}{1-e} \cdot \dfrac{mu}{m+2M}$, and that the particle will come to rest relatively to the scale-pan after a time $\dfrac{2eu}{g(1-e)}$.

17. Two balls, A and B, whose coefficient of elasticity is $\frac{1}{2}$, are moving with equal velocities in directions which make angles of 30° with their common tangent at the point of impact; compare their masses when the motion of A after impact is in the direction of that common tan-

gent, and find the distance between the balls 2 seconds after impact.

18. Find the velocity with which a perfectly elastic ball must be projected in a given direction from a point in the side AB of the square $ABCD$, so that after striking each of the sides in succession it may return to the point of projection, BC being vertical.

19. Two bodies are connected by an inextensible string which passes over a smooth fixed pulley, and are in motion. Prove that if weights be suddenly attached at the same instant to both the bodies they can be so arranged that there shall be no jerk of the string, and that the subsequent acceleration of the system will be in that case the same as before.

20. Two heavy bodies, P and Q, whose masses are m and m' respectively, are connected by an inextensible string which passes over a smooth fixed pulley. The heavier body P is perfectly elastic, and Q is inelastic; they start from rest at the same distance a above a fixed horizontal plane, and when P impinges on the plane and rebounds with unchanged velocity, Q strikes against a fixed obstacle and is reduced to instantaneous rest. Determine the subsequent motion, and show that the two bodies are again at instantaneous rest when P is at a height $\dfrac{m^2 a}{(m+m')^2}$ above the horizontal plane.

21. A, B, C are three points on the circumference of a circular ring fixed on a smooth horizontal table, and O is the centre. An imperfectly elastic ball is projected from A along AB and after rebounding at B and C returns to A. Determine the angle AOB.

22. The sides of a triangle ABC subtend equal angles at a point O within it. Prove that if from O three perfectly elastic balls be projected simultaneously with equal velocities in directions AO, BO, CO produced respectively, they will, after rebounding from the sides, all meet together simultaneously.

23. A particle is projected from a point at the foot of one of two parallel vertical smooth walls, so as after three

reflections at the walls to return to the point of projection, the last incidence being direct : prove that $e^3 + e^2 + e = 1$, and that the vertical heights of the three points of impact above the point of projection are as $e^2 : 1 - e^2 : 1$.

24. Two equal smooth balls, A and B, are lying very nearly in contact on a smooth horizontal table. A third ball, equal to either, impinges directly on A, the three centres lying in a straight line : prove that if e be greater than $3 - 2\sqrt{2}$, B's final velocity will bear to the initial velocity of the striking ball the ratio $(1 + e)^2 : 4$.

25. Two equal, smooth, and perfectly elastic balls, moving in directions at right angles to each other, impinge, their common normal at the instant of impact being inclined at any angle to the directions of motion : show that after impact the directions of motion will still be at right angles.

26. Two smooth elastic balls, moving in parallel directions, impinge on each other; show that if they are of equal mass their directions of motion will be turned through a right angle, if the inclination of their original paths to the line of impact be $\tan^{-1} \sqrt{e}$, where e is the coefficient of elasticity.

27. A ball A impinges obliquely on a ball B at rest; if the masses of A and B be m and m' respectively, and m be greater than em', show that the maximum deviation of A is

$$\tan^{-1} \frac{(1 + e) m'}{2 \sqrt{(m + m')(m - em')}}.$$

28. Two balls, whose masses are $2m$ and $3m$, are moving with the same velocity in directions making angles of 45° and 30° respectively with the common tangent at the point of impact; find the direction of motion and the velocity of their common centre of gravity after impact.

29. Two equal balls of elasticity e impinge, having before impact velocities u_1, v_1 in the direction of the common normal at the point of contact, and velocities u_2, v_2 perpendicular to this normal. If their motions after impact are in perpendicular directions, prove that

$$(u_1 + v_1)^2 + 4u_2 v_2 = e^2 (u_1 - v_1)^2.$$

30. Two equal and perfectly elastic spherical balls are projected in the same vertical plane from two points in the same horizontal line at a distance $g\sqrt{3}$ from each other; the former vertically with a velocity g, and the latter at an elevation of 30° with a velocity $2g$. Determine the motion of each after impact.

31. A perfectly elastic ball is thrown into a smooth cylindrical well from a point in the circumference of the circular mouth. Show that, if the ball be reflected any number of times from the surface of the cylinder, the intervals between the successive reflections will be equal.

If the ball be imperfectly elastic and be projected so as to pass through the axis of the cylinder, show that the intervals between the successive reflections form a series in Geometrical Progression.

32. In the last question, if the perfectly elastic ball be projected horizontally in a direction making an angle $\frac{\pi}{n}$ with the tangent at the point of projection, it will reach the surface of the water at the instant of the nth reflection, if the velocity of projection be that due to falling freely through a vertical space equal to $\frac{r^2}{d}\left(n \sin \frac{\pi}{n}\right)^2$, where r is the radius and d the depth of the well.

33. There are three equal and perfectly elastic balls A, B, and C. A is let fall from a given point; and at the moment when it reaches a given horizontal plane B is let fall from the same point; and at the moment when A in returning meets B, C is let fall. Show that B will meet C for the second time where it first met A.

34. ABC is an equilateral triangle; at A lies a ball, an equal ball strikes it driving it along AC, and itself passing through the middle point of BC. Show that the original direction of motion of this ball made with AC an angle $\tan^{-1}\frac{1-e}{2\sqrt{3}}$.

35. A, B, C are three equal smooth balls situated on a horizontal table at the angular points of an isosceles tri-

EXAMPLES. 235

angle having an obtuse angle at B. If A be struck so that having hit B it shall hit C, show that B will move in a direction inclined to AC at an angle ϕ, given by the equation

$$\sin 2\phi = \frac{3-e}{1+e}\sin 2A,$$

e being the coefficient of elasticity of the balls.

36. A heavy ball is thrown horizontally from A so as to hit a point B after one rebound from a horizontal plane C. Supposing e to be the coefficient of elasticity and the height of B from the plane to be e^2 times that of A, the height of A being such that a body would drop from it to the plane in 1 second, show that the point C where the ball must hit the plane divides the horizontal distance between A and B into two parts which are as $1 : e$.

37. A ball is projected from a point A at an elevation of 45° against a vertical wall BC, and in a vertical plane perpendicular to the wall; after impact at C it strikes the ground between A and B, and arrives at A after n rebounds. Find the ratio of BC to AB in terms of the coefficient of elasticity of the ball.

38. A parabola is placed with its axis vertical and vertex downwards. A perfectly elastic ball dropped vertically strikes the parabola with the velocity acquired in falling freely from rest through a space equal to one-fourth of the latus rectum; find where it must strike the parabola that after reflection it may pass through the vertex.

39. A particle is dropped from a point in a fixed circular hoop whose plane is vertical, the elasticity being perfect. Find the condition that after two rebounds it may rise vertically and determine in what ways this may happen.

40. A circular arc has its plane vertical. A perfectly elastic ball is projected from the arc along a horizontal diameter, and after one rebound at the arc returns to the point of projection. Show that the latera recta of the two parabolas described are as 4 to 1, and determine the velocity of projection.

236   EXAMPLES.

41. A number of balls whose elasticity is $\frac{1}{2}(\sqrt{2}-1)$ are let fall on an inclined plane, and each strikes it the second time twice as far down as it did the first time. Show that the points from which they fall lie in a plane perpendicular to the inclined plane, and intersecting it in a horizontal straight line.

42. A smooth sphere stands on a horizontal plane to which it is fixed, and from its highest point a perfectly elastic ball is projected in a direction inclined 45° to the vertical. Find the velocity of projection in order that the ball may strike the sphere once only, at an angular distance of 45° from the vertex, and prove that in that case the ball will strike the plane at a distance from the point of contact of the sphere equal to its diameter.

43. An elastic ball is projected from a point in a smooth inclined plane in the vertical plane containing the line of greatest slope on the plane. Find the condition that after three reflections it may return to the point of projection.

44. Each of two planes is inclined 45° to the horizon, and they intersect in a horizontal straight line; from any point in one of them it is possible to project a perfectly elastic ball in a plane perpendicular to the intersection of the planes, so as to return to the point of projection if the velocity of projection be not less than that acquired in sliding from the point of projection to the intersection of the planes.

45. Two vertical walls are inclined to one another at an acute angle a. A perfectly elastic ball projected horizontally from a point distant c from the ground and b from the intersection of the walls, comes to the ground, after striking both of them, at the same point as if it had fallen from rest. Find the direction of projection, and show that the space through which the ball would rise if projected vertically with the velocity of projection is equal to
$$\frac{(b \sin a)^2}{c}.$$

46. A series of balls of masses $M_1, M_2...$ are arranged

with their centres in a straight line, and the coefficient of elasticity between the rth and $(r+1)^{th}$ balls is $\dfrac{M_{r+1}}{M_r}$. Prove that if M_1 impinge directly upon M_2 at rest, and so on, the velocity of each ball between its two impacts will be equal to the initial velocity of M_1.

47. A very small heavy pan is supported by three strings passing over pulleys which are situated in a horizontal plane at the angular points of an equilateral triangle, the strings sustaining at the other ends equal weights hanging freely. If a given weight be dropped into the pan from a given height, find the velocity with which the pan will begin to descend.

48. Two equal buckets are connected by a string without weight passing over a smooth pulley, and over one of the buckets a heavy chain is held by its upper end, with its lower end just above the bottom of the bucket. If the upper end of the chain be let go, prove that the equilibrium may be maintained by pouring water *gently* and at a uniformly increasing rate into the other bucket, provided the weight of water which can be poured in is three times the weight of the chain.

After the chain has entirely fallen in, find the pressure on the bottom of the bucket in which it lies, supposing the flow of water then to cease.

49. Two equal perfectly elastic balls are let fall at the same instant from altitudes $\dfrac{1}{2}g$ and $\dfrac{9}{2}g$ respectively above a horizontal table, but not in the same vertical line; show that at the end of $6n \pm 1$ seconds the velocity of their centre of gravity suddenly changes from g to 0 or from 0 to g.

50. A ball A impinges on an equal ball B at rest; show that if the velocities after impact are equal, the change of direction in the motion of A is $\tan^{-1}\sqrt{e}$.

51. A horizontal circle ABC rests on a smooth table. A ball projected from A is reflected at B and C and returns

to A; show that the time from A to B is to that from C to A as e to 1.

52. A perfectly elastic ball is dropped from a point P and impinges on an inclined plane at Q. If PN be perpendicular to the plane, show that the range is equal to $8QN$, and hence find the locus of P in order that the particle may, after one reflection, pass through a fixed point on the plane.

53. Two weights are connected by a string which passes over a smooth fixed pulley, and the heavier rests on the ground; the lighter is raised a given height above its position of rest and then let go; show that they will make a series of jumps decreasing in Geometrical Progression, and find the common ratios, the lighter weight being supposed never to reach the ground.

54. A rough body, whose mass is 2 lbs., rests on a rough plane inclined 30° to the horizon, the coefficient of friction being $\frac{3}{4}$. An inelastic smooth body, whose mass is 1 lb., descends from a point on the plane distant 10 feet from the first body so as to impinge upon it directly, and the two slide on together. Find how far they will go before coming to rest.

55. Two smooth planes OA, OB each inclined to the horizon at the same angle a, which is less than $\frac{\pi}{4}$, intersect in a horizontal straight line. An inelastic ball descends from rest at A; show that the time which elapses before it is reduced to rest is to the time of descending AO as

$$\cot^2 a : 1.$$

56. Two elastic balls are moving in opposite directions before impact with velocities inversely proportional to their masses; compare their velocities after impact.

57. A perfectly elastic ball is projected with a given velocity from a point between two parallel vertical walls, and returns to the point of projection after being once reflected at each wall. Show that the angle of projection may be either of two complementary angles.

EXAMPLES. 239

58. A series of perfectly elastic balls are arranged in the same straight line; one of them impinges directly on the next, and so on; show that if their masses form a geometrical progression of which the common ratio is 2, their velocities after impact will form a geometrical progression of which the common ratio is $\frac{2}{3}$.

59. Two equal balls of radius a are in contact, and are struck simultaneously by a ball of radius c, moving in the direction of the common tangent to the first two balls at their point of contact; if all the balls be of the same material, the coefficient of elasticity being e, find the velocities of the balls after impact, and prove that the impinging ball will be reduced to rest if

$$2e = \frac{c^2(a+c)^2}{a^3(2a+c)}.$$

60. A smooth inelastic ball slides from rest down a length (l) of a plane inclined 30° to the vertical, and impinges on a horizontal rail, parallel to the plane and at a distance from it equal to one-half the radius of the ball. Neglecting the thickness of the rail, prove that the ball will afterwards strike the plane at a distance $3l$ from its point of contact when striking the rail.

61. An inelastic ball of mass m lies on a horizontal plane; another inelastic ball of mass m' falls vertically and strikes it in such a manner that the line joining their centres at the moment of impact makes an angle a with the vertical. Show that the direction of motion of the second ball immediately after impact will make an angle θ with the vertical determined by the equation

$$\tan \theta = \frac{m}{m+m'} \cot a.$$

62. A ball of elasticity e is projected from a given point A, with velocity V, so as to strike a vertical wall distant a feet from A, and after impact to strike the horizontal plane through A at a point B, distant b feet from the wall. If c be the perpendicular distance of A from the line drawn through B at right angles to the wall,

show that the least possible value of V is given by the equation

$$V^2 = \frac{g}{e}\sqrt{c^2e^2 + (b+ae)^2}.$$

63. A number of particles are let fall from the directrix upon the convex arc of a parabola whose axis is vertical and latus rectum equal to $4a$. Show that the parabolas which they describe after impact upon the curve have a common directrix at a distance $a(1-e^2)$ below that of the fixed parabola, where e is the coefficient of elasticity between the particles and curve.

64. Two equal particles are projected simultaneously from different points. If they impinge and after impact move vertically, prove that $\tan\theta - \tan\phi$ is constant, where θ and ϕ are the inclinations to the horizon of the directions of motion of the particles at any, the same, moment previous to their impact.

65. An elastic sphere is at rest on a plane. The plane and sphere are simultaneously hit by another smooth sphere, whose coefficients of elasticity with the first sphere and plane are the same and equal to e. Determine e when the directions of motion of the second sphere, after and before impact, are equally inclined to the plane.

66. A very small elastic ball is projected with a given velocity from one extremity of a diameter of a horizontal circular hoop, which rests on a smooth horizontal table, and after reflection at the curve passes through the other extremity of the diameter. Find the coefficient of elasticity in order that the whole time occupied in the motion may be n times that of describing the diameter with the initial velocity; and the greatest and least values n can have.

67. A ball of given mass lies touching a smooth wall. Another moving at right angles to the wall impinges on it obliquely. The balls being inelastic, find their velocities immediately after impact.

68. An imperfectly elastic ball is projected from a given point in a horizontal plane against a smooth vertical wall

in a direction making a given angle with the vertical: find where it strikes the horizontal plane, and prove that the locus of these points for different vertical planes of projection is an ellipse.

69. A heavy chain hangs vertically from its upper end with its lower end just in contact with a smooth plane inclined at an angle a to the horizon: if the string be allowed to fall, find the pressure on the plane at any time.

70. Two smooth equal balls are placed in contact on a smooth table; a third equal ball strikes them simultaneously and remains at rest after the impact; show that the coefficient of restitution is $\dfrac{2}{3}$.

71. A uniform flexible chain of indefinite length, the mass of the unit of length of which is m, lies coiled on the ground while another portion of the same chain forms a coil on a platform at a height h above the ground, the intermediate portion passing round the barrel of a windlass placed above the second coil. An engine which can do H units of work per the unit of time is employed to wind up the chain from the ground and to let it fall into the upper coil. Show that the velocity of the chain can never exceed the value of v determined from the equation

$$mghv + mv^3 = H.$$

72. A, B, C are three perfectly elastic balls of equal mass lying on a horizontal plane. If A and B are connected by a tight inelastic string, and C is projected so as to strike A directly with velocity V, prove that C will rebound with velocity

$$V \frac{\cos^2 \theta}{3 + \sin^2 \theta},$$

where θ is equal to the angle BAC and is less than a right angle.

73. A bullet is fired in the direction towards a second equal bullet, which is let fall at the same instant. Prove that the two bullets will meet, and that if they coalesce the latus rectum of their joint path will be one quarter of the latus rectum of the original path of the first bullet.

74. A bucket and a counterpoise, connected by an inelastic string passing over a pulley, just balance one another, and an elastic ball is dropped into the centre of the bucket from a distance h above it; find the time that elapses before the ball ceases to rebound, and prove that the whole descent of the bucket during this interval is $\dfrac{4mh}{2M+m} \dfrac{e}{(1-e)^2}$, where m, M are the masses of the ball and bucket and e is the coefficient of restitution.

75. Two equal particles A, B, of imperfect elasticity e, move with equal uniform velocity in the same straight line. B impinges perpendicularly on a wall. Show that there will always be two impacts between A and B, and two between B and the wall, and that if there is a third collision between the balls,
$$e < 2 - \sqrt{3}.$$

76. A particle is projected from a point in a smooth plane inclined at an angle a to the horizon, in a vertical plane which cuts the inclined plane in a horizontal line, and at an angle θ to the horizon. Prove that after n rebounds the space traversed in the direction of the line of greatest slope on the inclined plane is
$$a \sin a \tan \theta \cdot \frac{e(1 - e^{n-1})}{1 - e},$$
where a is the horizontal space described, and e the coefficient of restitution.

77. A ball having descended to the lowest point of a circle through an arc whose chord is C drives an equal ball up an arc whose chord is c: show that the common elasticity (e) of the two balls is given by the relation
$$1 : e :: C : 2c - C.$$

78. An elastic ball being projected at any elevation is continually reflected from a horizontal plane, and the sum of the areas of all the parabolas described : area of the first parabola :: 8 : 7. Find the elasticity of the ball.

CHAPTER V.

MISCELLANEOUS.

245. *A particle falls down a smooth curved tube; it is required to find its velocity at any point of the tube.*

Since the tube is smooth the only force which it exerts upon the particle is at right angles to the direction of the particle's motion. Hence no work is done upon the particle by the action of the tube, and the only force which does work upon the particle is its weight, so that the kinetic energy generated must be the equivalent of the work so done, and therefore depend simply on the *vertical* height through which the particle has fallen (Art. 171). Its velocity is therefore the same as if it had fallen *vertically* through the same difference of level. Hence, if a particle starting with velocity u move along a smooth tube (or other surface) through a *vertical* distance h, its velocity at the end of the distance will be $\sqrt{u^2 + 2gh}$. The same will be true if the particle be constrained by an inextensible string so that it moves always at right angles to the string.

If the particle start from rest at A its velocity at any point, whose vertical depth below A is h, will be $\sqrt{2gh}$.

246. Similarly if a particle be projected up a smooth tube with velocity u, its velocity after rising through a vertical height h will be $\sqrt{u^2 - 2gh}$. If h be the greatest height to which it will rise, we must have $u^2 = 2gh$, or $h = \dfrac{u^2}{2g}$. Hence the particle will rise up the tube to the same vertical height to which it would rise if it were free and projected vertically upwards with velocity u.

247. If the tube be bent, as in the subjoined figure, and the particle fall down one arm from a height h above the

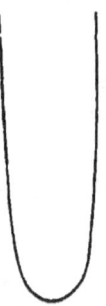

lowest point, it will rise up the other arm to the same height, for its velocity at the lowest point will be $\sqrt{2gh}$, and this will just carry it to a height h up the other arm. The particle will afterwards descend again, and will continue to oscillate to a height h on each side of the lowest point of the tube.

248. If a particle move subject to any constraints whatever, such that the force exerted upon the particle by the constraint is always perpendicular to the direction in which it is moving, no work is done upon or against the particle by the means of constraint. Hence the change in the kinetic energy of the particle produced during its motion from one point to another must be equivalent to the work done upon it by the forces to which it is subject, the action of the constraints being left out of consideration. Hence, if a particle be moving under the action of gravity but constrained by any smooth surfaces, inextensible strings, or system of frictionless link-work, the change in its kinetic energy will depend only on the vertical distance through which it has risen or fallen.

We have already, under the head of projectiles, considered a case in which the force acting upon a moveable particle is inclined to the direction of motion, and we found that in this case the kinetic energy of the moving particle depended only on its vertical distance below the directrix of the parabola which it described. The motion of the earth about the sun affords another example of the motion of a body under the action of a *conservative* force but subject to no constraints, and the velocity of the earth in its orbit depends only on its distance from the sun.

249. As an illustration of the preceding articles we will take the following example.

A number of heavy particles slide from rest at the vertex down a smooth tube in the form of a parabola whose axis is vertical, and are allowed to quit the tube at different points. Find the locus of the foci of the trajectories subsequently described by them.

Let S be the focus of the parabolic tube, ZX its directrix, AL the tangent at the vertex. Suppose a particle to quit the tube at any point P. Draw PKH parallel to the axis. Let PT be the tangent at P to the curve of the tube. Then, since the particle starts from rest at A, its velocity at P is that due to falling freely from the point K. AK must therefore be the directrix of the parabola subsequently described by the particle (see Art. 187),

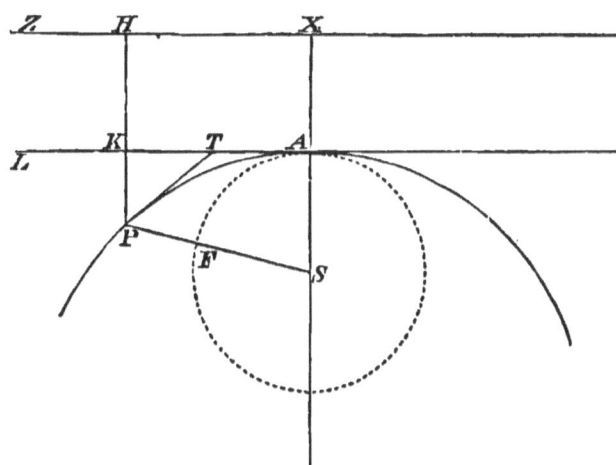

and since PT is the direction of the particle's motion at P, it is a tangent to this parabola. Now, since PK is the perpendicular on the directrix and PT the tangent at P, if F be the focus of the parabola, the angle FPT is equal to TPK. Hence F lies in SP. Also FP is equal to PK

and SP to PH. Hence SF is equal to HK, that is, to SA. The locus of the foci is therefore a circle whose centre is S and which passes through A.

Since SF is equal to SA, S is a point on the trajectory described by the particle which leaves the tube at P. Hence S is a point on the parabolas described by each of the particles after leaving the tube. These parabolas are in fact the same as would be described by a series of particles projected in different directions from S, each with the velocity which it would acquire in falling freely from A to S, and the curve of the tube is the envelope of all these parabolas. (See Art. 198.)

250. In Art. 245, we have supposed the particle to be constrained to move in a smooth tube, but the proof there given will be equally true (Art. 248) if the constraint be produced by any other means, provided it exert no force upon the particle in the direction of its motion or in the opposite direction. For example, if a particle slide down a *smooth* surface of any form whatever, or if it be fastened to one end of a string of *constant length* (and whose mass may be neglected), the other end of the string being attached to a fixed point. In all these cases the change of the velocity of the particle will depend only upon the vertical height through which it has fallen.

251. We are now in a position to understand the method by which Newton arrived at the law of impact, enunciated in Art. 207, and the determination of the coefficient of elasticity for different substances from the results of experiment.

A and B are two spherical balls suspended by strings from fixed points so that the centres of both are free to move in the same vertical plane. Let this plane be that of the paper. Now, if the diameters of the balls are small compared with the length of the strings, we may, without introducing any considerable error, suppose them to move as particles situated at their centres of gravity. The length of each string and its point of suspension are carefully adjusted, so that when at rest the balls may be just in contact and the line joining their centres may be

NEWTON'S EXPERIMENT ON IMPACT. 247

horizontal and in the vertical plane in which they move. Let a, b denote their centres in this position.

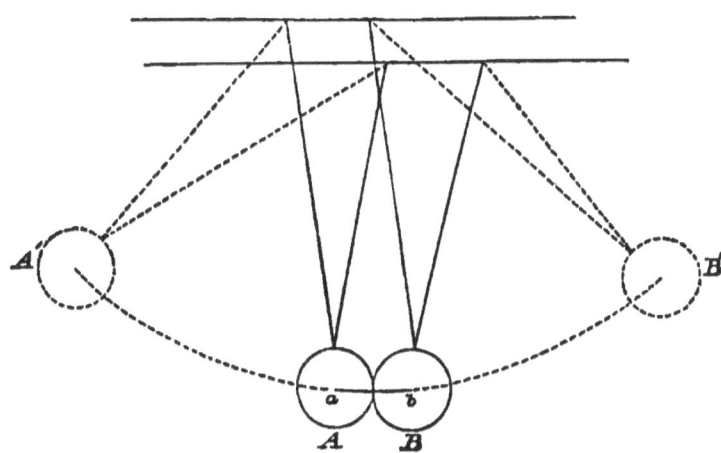

Let the centre of the ball B be raised to B', its strings being kept tight, and then let B be allowed to fall from rest. Let the vertical height of B' above the line ab be h feet. Then the velocity of the centre of the ball when it strikes A being denoted by u we have

$$u^2 = 2gh;$$
$$\therefore u = \sqrt{2gh},$$

and this is the velocity of B relative to A just before impact. Also, since at the instant of impact B is at its lowest position, its centre will be moving horizontally, that is, along ba. After impact, the centre of the ball A will move along the arc of a vertical circle. Let A' be the extreme point which it reaches, and let the vertical height of A' above the line ab be denoted by k. Then, since the ball comes to rest at A', the velocity v with which it left a must be given by the equation

or
$$v^2 = 2gk,$$
$$v = \sqrt{2gk}.$$

Again, after impact, the ball B will, on leaving A, either return towards B' or continue moving along its circle in the same direction as before, but with a diminished velocity; or it may, as a particular case, come to rest at once. Suppose it to return towards B' and its centre to rise to a vertical height h' above the line ab. Let v' be its velocity immediately after impact with A. Then, since the greatest height to which it rises above ab is h', we must have

or
$$v'^2 = 2gh',$$
$$v' = \sqrt{2gh'}.$$

Hence the velocity of the centre of B relative to that of A immediately after impact, that is, $-(v+v')$, is numerically equal to $-\sqrt{2g}\,(\sqrt{k}+\sqrt{h'})$.

Now Newton found by measuring the heights h, k and h' that so long as the materials of which the balls were composed were the same the ratio

$$\frac{\sqrt{k}+\sqrt{h'}}{\sqrt{h}}$$

was always constant, whatever were the relative dimensions of the balls or the height h to which B was raised. But this ratio, with a negative sign prefixed, is the ratio of the velocity of B relative to A after and before impact. Hence this latter ratio is constant. We have called the ratio $\dfrac{\sqrt{k}+\sqrt{h'}}{\sqrt{h}}$ the coefficient of elasticity, and have denoted it by e. It is always less than 1.

252. We have said that after impact upon A the ball B may return towards B', may come to rest, or may go on in the same direction as before, but with diminished velocity. Its behaviour in this respect will be determined by the value of e, and the ratio of the masses of the balls. If it come at once to rest, h' is zero and e becomes $\dfrac{\sqrt{k}}{\sqrt{h}}$. If it proceed in the same direction as before, but with velocity v'', then its velocity relative to A immediately after impact is $-(v-v'')$; and if h'' be the height to which it rises, we have for the ratio of the relative velocities, after and

THE CYCLOID. 249

before impact, the expression $\dfrac{\sqrt{k}-\sqrt{h''}}{\sqrt{h}}$, and this expression will be found to be constant. The coefficient of elasticity will in this case be $\dfrac{\sqrt{k}-\sqrt{h''}}{\sqrt{h}}$

If the balls be inelastic they will proceed after impact with the same velocity, and will therefore rise to the same height. Hence $k=h''$ and $\dfrac{\sqrt{k}-\sqrt{h''}}{\sqrt{h}}$ becomes zero, as of course it should. No known bodies are, however, perfectly inelastic.

253. We propose now to investigate the motion of a particle constrained to move under the action of gravity upon a smooth cycloid whose axis is vertical and vertex downwards. Before doing this we must examine some of the properties of the cycloid. The proofs given in the following articles are due to Dr. W. H. Besant, of St. John's College.

DEF. *A cycloid is the curve generated by a point in the circumference of a circle, while the circle rolls (without slipping) along a straight line.*

Suppose PQK to be the circle, Q the point fixed in its circumference, and the circle to roll along the under side

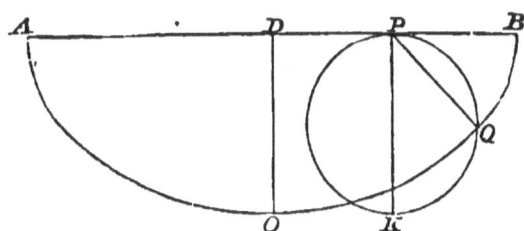

of the straight line AB, starting from the position in which Q is in contact with the line at B. Then Q will generate the cycloid BOA. If O be the position of Q when the diameter through Q is perpendicular to AB, then O is called the vertex of the cycloid, and OD, the diameter of the generating circle which passes through O, is called its axis. The points A and B are the cusps of the curve, and the line AB its base. If the diameter of the generating circle be denoted by a, then OD is equal to a, and AB to the circumference of the circle, that is, to πa. The length a of the diameter of the generating circle is called the parameter of the cycloid.

254. Let Q be any point on the cycloid, and P the point at which the generating circle touches the base when

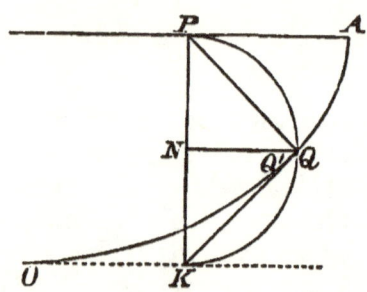

the tracing point is at Q. Let Q' be the position of the tracing point when the circle has turned through the indefinitely small angle ϕ from this position. Then as the circle *begins* to roll, it turns about the point P as an instantaneous centre of rotation, and ϕ being the angle through which it turns,

$$QQ' = PQ \cdot \phi.$$

Now if a particle starting from rest at A fall down the arc of the cycloid, which we suppose smooth, its velocity at Q will be given by the equation

$$v = \sqrt{2g \cdot PN}.$$

If we suppose this velocity to remain constant while the particle moves over the indefinitely small arc QQ', then the time taken to pass from Q to Q' will be denoted by $\dfrac{QQ'}{\sqrt{2g \cdot PN}}$. Call this time τ, then

$$\tau = \frac{QQ'}{\sqrt{2g \cdot PN}} = \frac{PQ}{\sqrt{2g \cdot PN}} \cdot \phi.$$

THE ISOCHRONISM OF THE CYCLOID. 251

By similar triangles
$$PK : PQ :: PQ : PN,$$
or
$$PQ^2 = aPN;$$

$$\therefore \tau = \sqrt{\frac{a}{2g}} \cdot \phi;$$

and this is true for each indefinitely small arc into which AO may be divided. Hence the time in which the particle will descend from A to O is $\pi\sqrt{\dfrac{a}{2g}}$, since the circle turns through two right angles while the tracing point passes from A to O.

255. *A particle starts from rest at any point in the arc of a smooth cycloid whose axis is vertical and vertex downwards; to find the time of descent to the vertex.*

Let T be the point from which the particle starts. Through T draw TC parallel to AD, and let a second

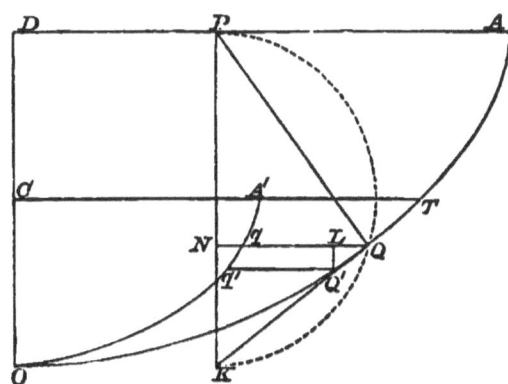

cycloid OA' be drawn, having its vertex at O and OC for axis. Let OD be denoted by a, OC by a'.

Let Q be any point on the first cycloid between T and O, Q' a contiguous point. Draw Qq, $Q'q'$ parallel to AD, meeting the second cycloid in q, q' respectively, and let Qq meet PK in N. Draw $Q'L$ perpendicular to Qq. Then

QQ' is ultimately perpendicular to PQ, that is, QQ' coincides with QK when Q' is indefinitely near to Q. Hence

$$QQ' : Q'L :: KQ : KN,$$

$$\text{or } \frac{QQ'}{Q'L} = \frac{KQ}{KN}.$$

But $\quad KN : KQ :: KQ : PK;$

$$\therefore \frac{KQ}{KN} = \sqrt{\frac{PK}{KN}};$$

$$\therefore \frac{QQ'}{Q'L} = \sqrt{\frac{PK}{KN}} = \sqrt{\frac{a}{KN}}.$$

Similarly $\quad \dfrac{qq'}{Q'L} = \sqrt{\dfrac{a'}{KN}};$

$$\therefore QQ' : qq' :: \sqrt{a} : \sqrt{a'}.$$

Now if a particle slide from rest at A' down the arc of the second cycloid, its velocity at q is the same as the velocity at Q of the particle which slides from T, since the vertical height through which each will have fallen is the same. Hence the time taken by the second particle to slide down qq' is to that taken by the first to slide down QQ' as arc qq' is to arc QQ', that is, as $\sqrt{a'}$ to $\sqrt{a}$: and this is true for each pair of corresponding elements into which the arcs $A'O$, TO can be divided. Therefore the time taken by the first particle to slide down TO is to that taken by the second particle to slide down $A'O$ as $\sqrt{a}$ to $\sqrt{a'}$. But the time taken by the second particle to slide down $A'O$ is, by the preceding article, $\pi\sqrt{\dfrac{a'}{2g}}$. Therefore the time taken by the first to slide from T to O is equal to $\pi\sqrt{\dfrac{a}{2g}}$, or the time from T to O is the same as from A to O. Hence the time taken by a particle to fall from rest at *any* point of the cycloid to the vertex is the same. This property is called the "*isochronism of the cycloid,*"

LENGTH OF THE ARC OF A CYCLOID. 253

The particle after passing O will then ascend the cycloid to the same height as the point T from which it has fallen. If we denote by T' the point at which it comes to rest, the time from T to T' is

$$2\pi \sqrt{\frac{a}{2g}}, \text{ or } \pi \sqrt{\frac{2a}{g}}.$$

The particle will then return from T' through O to T, and the whole time occupied in a complete oscillation or "swing-swang," that is, in passing from T to T' and back again, is

$$2\pi \sqrt{\frac{2a}{g}},$$

and is constant, however great or small the arc of vibration may be, provided the particle do not leave the curve.

256. Let Q be any point on the cycloid, and PK the corresponding vertical diameter of the circle. Suppose the

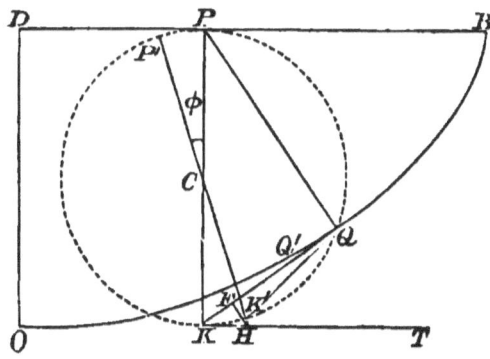

circle to roll through the indefinitely small angle ϕ, and Q thereby to move to Q'. Let $P'K'$ be the diameter which then becomes vertical. Then the angle $P'CP$ is equal to ϕ. Join QK'.

Draw KT parallel to DB and let $P'K'$ meet KT in H. Draw HF perpendicular to QK. Then QF is ultimately equal to QK', and $KF = QK - QK'$.

Now when ϕ is indefinitely small, KH is equal to KK', that is, to $CK \cdot \phi$. Hence

$$KH = \frac{a}{2} \cdot \phi.$$

Also
$$KF = KH \cdot \cos QKT$$
$$= KH \cos KPQ \,;$$

$$\therefore KF = \frac{a \cos KPQ}{2} \phi$$

$$= \frac{PQ}{2} \cdot \phi$$

$$= \frac{1}{2} QQ',$$

or $QQ' = 2 (QK - QK')$.

Hence as the circle rolls through any very small angle the increment QQ' of the arc of the cycloid is equal to twice the decrement of the chord QK; and this being always true up to the time when Q coincides with O, in which case both the arc QO and the chord QK vanish together, it follows that the arc QO is equal to twice the chord QK.

Hence also the arc BO is equal to $2PK$, that is, to $2a$, and the length of the whole arc of the cycloid is $4a$.

257. Let $BQOA$ be a cycloid whose base is AB and vertex O. Let the axis OD be vertical and be represented by a. Then a is the parameter of the cycloid. Produce OD to X making DX equal to OD. Draw two semicycloids having their vertices at A and B respectively, and each having a cusp at X. Then the parameter of each of these semicycloids is a. Draw XY parallel to AB, and BY perpendicular to it. Let PQK be any position of the generating circle of the first cycloid, Q the corresponding point on the curve. Produce KP to meet XY in K'. On PK' describe a circle, and produce QP to meet this circle in R. Join $K'R$.

THE CYCLOIDAL PENDULUM.

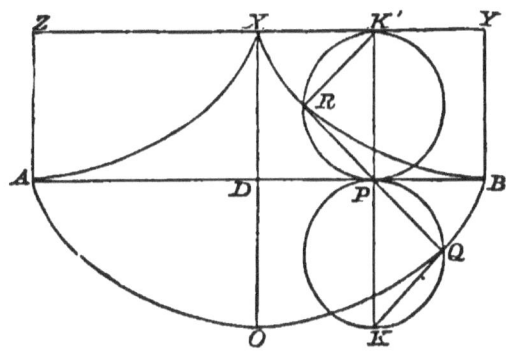

Then $\quad XY = DB = \dfrac{\pi}{2} a = \text{arc } PQK.$

Also $\quad K''Y = PB = \text{arc } PQ,$

and $\quad K''X = XY - K''Y;$

$\therefore K''X = \text{arc } KQ.$

But the triangles PQK, PRK'' are equal, and therefore QK is equal to RK''. Therefore

$$\text{arc } QK = \text{arc } K''R.$$

Hence $\quad K''X = \text{arc } K''R,$

and R is therefore a point on the cycloid XB. Also by the preceding article the arc RB is equal to twice the chord RP, that is, to RQ.

Again, since as R traces out the curve XB, K'' is the instantaneous centre about which $K''R$ is turning, $K''R$ is the normal at R to the cycloid XB, and RP is the tangent to the same. Hence if a string be fastened at X, and wrapped tightly round the cycloid XB, and a particle be attached to it at B, then if the system be left free to move, the string remaining tight, the particle will describe the semicycloid BO. Then as the string wraps up on the semicycloid XA, the particle will describe the semicycloid OA. Hence by this means a particle may be made to oscillate in a cycloid.

If the mass of the string be insensible compared with the mass of the particle, the arrangement is called a simple pendulum.

If l be the length of the string, then $l = 2a$, and the time in which the particle oscillates from its highest position on one side of O to that on the other side of O is, by Art. 255,

$$\pi \sqrt{\frac{l}{g}},$$

and is independent of the amplitude of the vibration. The time of a complete oscillation is

$$2\pi \sqrt{\frac{l}{g}}.$$

258. When the particle is very near to O, the string by which it is suspended will be very nearly straight. Hence if a circle be described having X for centre and XO for radius, it will very nearly coincide with the cycloid at points near to O. Hence if a particle oscillate in this circle through any small distances on each side of O, its motion will be nearly the same as if it moved on the cycloid, and the time of a semi-oscillation, that is, from the extreme point on one side of O to the extreme point on the other side of O, will be very nearly $\pi \sqrt{\frac{l}{g}}$.

If the arc of oscillation be more than 3 or 4 degrees a considerable difference exists between the time in the circle and in the cycloid, the oscillations in the circle being slower the greater the amplitude.

A simple pendulum as described above is of course a pure conception, and can never be realized experimentally. If, however, any rigid body be made to oscillate through a very small angle in one plane about a fixed point under the action of gravity, and the form and dimensions of the body be known as well as its density at every point, we can calculate the length of a simple pendulum which will oscillate in the same time; but this calculation requires the methods of Rigid Dynamics. The simple pendulum which performs its vibrations in the same time as any rigid body or compound pendulum, is called the *simple equivalent pendulum*.

The method of suspension explained in the preceding article is sometimes adopted in clocks. The pendulum is supported by a short steel spring, which, as the pendulum oscillates, is made to wrap itself on two cycloidal cheeks, like the string in the last article. As the arc of vibration of the pendulum is never more than a few degrees in amplitude, only a very small portion of the cycloid is required for either guiding cheek.

Suppose we observe the time in which any known pendulum performs a large number (say 200 or 300) of very small oscillations. Then, dividing this time by the number of oscillations, we can find the time occupied by each. Let t seconds be the time occupied in performing a semi-oscillation; then, if l be the length of the simple equivalent pendulum, $t = \pi \sqrt{\dfrac{l}{g}}$

Hence, l being known, we can calculate the value of g. It is from experiments of this description that the most exact values of g have been determined.

259. As an illustration of the preceding articles we will determine the length of a simple pendulum which will perform a semi-oscillation in one second in London, the value of g being supposed equal to 32·19.

A pendulum which performs a semi-oscillation in one second is called a seconds' pendulum. By "beats" of a pendulum are always meant semi-oscillations.

If l be the length of the seconds' pendulum in feet, we have

$$\pi \sqrt{\dfrac{l}{g}} = 1 \,;$$

$$\therefore\; l = \dfrac{g}{\pi^2}$$

$$= \dfrac{32 \cdot 19}{3 \cdot 1416^2}$$

$$= 3 \cdot 262 \ldots$$

or the length of the seconds' pendulum is about 39·144 inches.

260. The reason of the isochronism of the pendulum is simply that the acceleration with which it moves is always proportional to its distance (measured along the path of the bob) from its position of rest. If this condition be fulfilled, it matters not whether the body be moving in a straight or curved line, it will still be isochronous in its vibrations, and the solution of the whole problem will be exactly similar to that of the pendulum. Hence, if a particle of mass M be free to move in any path under the action of a force along its path towards some point O and always equal to $\mu M d$, where d represents the distance of the particle from the point O measured along the path, the particle will perform isochronous vibrations about O, whose period will be $\dfrac{2\pi}{\sqrt{\mu}}$, no matter what may be the amplitude of the oscillation.

Similarly, if a rigid body be capable of turning about a fixed axis, and it can be shown that its angular acceleration towards its position of rest is always proportional to the angle through which it has been deflected, the body will oscillate about its position of rest according to the same law as the pendulum, its time of oscillation being independent of the amplitude. If the angular acceleration towards the position of rest be $\mu\theta$, where θ represents the angle through which the body has been deflected from rest, the period of a complete vibration will be $\dfrac{2\pi}{\sqrt{\mu}}$.

Any sounding body which is emitting a pure tone executes its vibrations in accordance with the same law as the pendulum, the acceleration of each point being always proportional to its distance from its position of rest, and such oscillations are, therefore, frequently called harmonic vibrations.

261. Suppose a particle to be describing the curve APB, and let v be its velocity at P; then, if PT be the tangent at P, the direction of its motion at P is along PT. Let Q be a point on the curve very near to P, v' the

NORMAL ACCELERATION. 259

velocity of the particle at Q, QT' the tangent at Q, and let the normals at P and Q intersect in O. Then, when

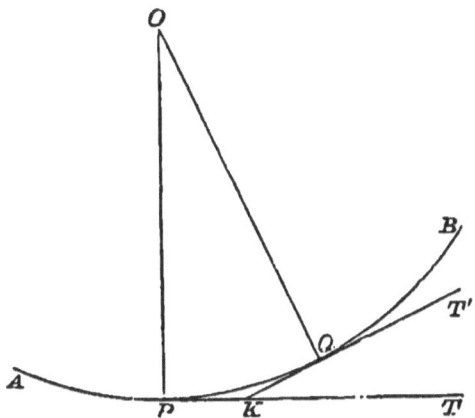

Q is indefinitely near to P, O is the centre of curvature of the curve APB at P. Let the angle POQ be denoted by θ, then $T'KT$ is equal to θ.

Now the velocity of the particle when at P is entirely along PT; its velocity parallel to OP is therefore zero. The velocity of the particle when at Q is v' along QT', and its velocity parallel to PO is therefore $v' \sin \theta$. Now the time occupied by the particle in moving from P to Q lies between $\dfrac{PQ}{v}$ and $\dfrac{PQ}{v'}$, and if PQ be indefinitely small, we may take v equal v', and the time from P to Q becomes $\dfrac{PQ}{v}$. Hence during the time $\dfrac{PQ}{v}$, a velocity represented by $v \sin \theta$ is generated in the particle in a direction parallel to PO. The measure of the acceleration which will generate this velocity in the time $\dfrac{PQ}{v}$ is $v \sin \theta \cdot \dfrac{v}{PQ}$, or $\dfrac{v^2 \sin \theta}{PQ}$. Now

$$\frac{v^2 \sin \theta}{PQ} = v^2 \frac{\theta}{PQ} \cdot \frac{\sin \theta}{\theta}.$$

Hence the particle must, while passing from P to Q, be

moving with an acceleration whose measure is the limit of the expression $v^2 \dfrac{\theta}{PQ} \cdot \dfrac{\sin \theta}{\theta}$, when θ is indefinitely diminished. But the limit of $\dfrac{\sin \theta}{\theta}$, when θ is indefinitely small, is unity, and the limit of $\dfrac{PQ}{\theta}$ is PO, that is, the radius of curvature at P. Let this be denoted by ρ. Then the acceleration of the particle at P in the direction PO is measured by $\dfrac{v^2}{\rho}$.

Hence, if the mass of the particle be denoted by m, it must be acted upon by a force in the direction PO represented by $\dfrac{mv^2}{\rho}$.

If the velocity remain constant, the particle has no acceleration in the direction of motion. Hence the resultant force upon it is a force $\dfrac{mv^2}{\rho}$ acting inwards along the normal at P.

If the curve described be a circle of radius r, then ρ is equal to r, and the particle is always acted upon by a force $\dfrac{mv^2}{r}$ towards the centre of the circle.

In the case of motion in a circle, if ω denote the angular velocity about the centre, $v = \omega r$ and the force towards the centre becomes $m\omega^2 r$.

If the particle make n complete revolutions per second, $\omega = 2\pi n$ and the force towards the centre becomes $mr(4\pi^2 n^2)$.

This method of determining the acceleration of a point along the normal to the curve in which it is moving is due to Dr. Besant.

262. In the case of a particle being prevented from leaving the circle by a string attached to the centre, this force is supplied by the tension of the string. Hence the string must exert a force upon the particle represented by

CONICAL PENDULUM.

$m\dfrac{v^2}{r}$, and since action and reaction are equal and opposite, it follows that the particle exerts a force upon the string acting from the centre of the circle, and also represented by $m\dfrac{v^2}{r}$. This action of the particle upon the string or other means of constraint is frequently called centrifugal force. It should always be borne in mind that the force acting upon the particle is towards the centre of the circle, but that the action of the particle upon its means of constraint is in the opposite direction, and is properly termed centrifugal force.

263. As an example of the preceding article we may take the following. Suppose a particle P, of mass m, to be attached to one end of a string of length l, the other end of which is fixed at A. The particle is made to describe a horizontal circle with uniform velocity, such that it makes n complete revolutions per second. It is required to find the inclination, θ, of the string to the vertical, and the tension of the string.

Let O be the centre of the circle described by the

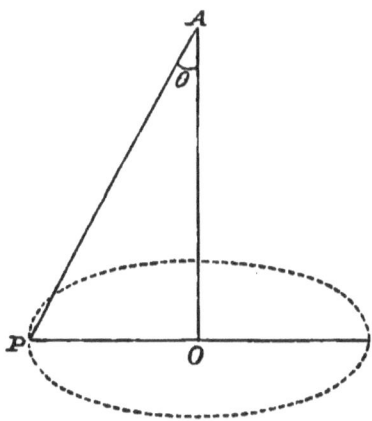

particle. The velocity of the particle is $2\pi n \cdot OP$. The acceleration towards O is $(2\pi n)^2 \cdot OP$, and the resultant force which must act on the particle to produce this acceleration is

$$m \cdot (2\pi n)^2 \cdot OP,$$

CONICAL PENDULUM.

This is, therefore, equal to the horizontal component of the tension in AP. The vertical component must balance the weight of the particle, and therefore be equal to mg. But AO is to OP as the vertical component of the tension in AP is to the horizontal component.

$$\therefore AO : OP :: mg : m(2\pi n)^2 OP;$$

$$\therefore AO = \frac{g}{(2\pi n)^2} \text{ or } \frac{g}{\omega^2}.$$

The vertical height therefore of the point of suspension, A, above the particle is equal to $\frac{g}{\omega^2}$, and depends only on the value of "g" and on the angular velocity of the particle. This result is of great importance.

The tension, T, in AP may be determined at once, for its vertical component is equal to mg, the weight of the particle,

$$\therefore T : mg :: AP : AO$$

$$:: AP : \frac{g}{\omega^2};$$

$$\therefore T = m\omega^2 . AP = m\omega^2 l.$$

Or we may proceed thus:

The tension in AP is to its horizontal component as AP to OP.

$$\therefore T : m\omega^2 OP :: AP : OP;$$

$$\therefore T = m\omega^2 AP$$

$$= m\omega^2 l.$$

To determine the angle θ we have

$$\cos \theta = \frac{AO}{AP}$$

$$= \frac{mg}{T}$$

$$= \frac{mg}{m\omega^2 l}$$

$$= \frac{g}{\omega^2 l}.$$

EXAMPLES OF CENTRIFUGAL FORCE. 263

or
$$\frac{g}{4n^2\pi^2 l}.$$

For example, suppose l equal to 2 feet and m to be 20 pounds, and that the system makes 10 revolutions per second, then taking g equal to 32, we have

$$T = 20 \cdot 20^2 \pi^2 \cdot 2$$
$$= 16000\pi^2,$$

or the tension of the string is $16000\pi^2$ dynamical units of force, that is, equal to the weight of $500\pi^2$ pounds.

If the string in this example be replaced by a rigid rod, which can turn about A in a ball and socket joint, we obtain the instrument known as a conical pendulum.

263a. If a number of particles be suspended by strings or thin rods of unequal lengths from the same point A, and all be made to execute the same number of revolutions per minute, all the particles will revolve in the same horizontal plane at the depth $\frac{g}{\omega^2}$ below A, and if the particles are equal the tensions of the several rods or strings will be proportional to their respective lengths. If the length of one of the pendulums is less than $\frac{g}{\omega^2}$ that pendulum will hang vertically and simply rotate about its own axis.

263β. The pendulum bob, for in the case of a spherical bob the centre of the ball will behave in the same way as the particle considered above, may be supported on a bar bent to the form of any curve instead of being suspended by a string or rod. If the bar be smooth, its pressure on the ball must be along its normal, and we have therefore simply to replace the string AP, of Art. 264, by the normal to the curve of the bar. If the bar be bent into such a curve that the subnormal, which corresponds to AO, is constant and equal to $\frac{g}{\omega^2}$, the bob will be in relative equilibrium upon the bar in every position. If the angular velocity be ever so little less than ω the bob will sink to the lowest position on the bar. If it be ever so little greater than ω the bob will rise to the highest possible position, and such an instrument will be infinitely sensitive in recording any change of angular velocity from the standard velocity ω. The curve whose subnormal is constant is a parabola, and the bar must therefore be bent into the form of a parabola, with its axis vertical and coinciding with the axis of rotation, and its latus rectum must be $\dfrac{2g}{\omega^2}$.

If the balls be made to rotate by being connected with the crank shaft of a steam-engine, and be connected by suitable mechanism with the steam supply, so that the steam is gradually cut off as the balls fly out, we have the so-called "parabolic governor," the principal fault of which is that it is too sensitive and is liable to cause, by the completeness with which it cuts off the steam when the speed is only slightly increased, greater fluctuations of speed than those it is designed to prevent. An approximation to a parabolic governor is frequently obtained by suspending the balls by rods from two points at the extremities of a horizontal cross arm, so that the points of suspension are at some distance from the axis of rotation. The rods are provided with slots, and are made to cross one another and the axis. The centre of suspension of each ball being on the opposite side of the axis from the ball itself, may be regarded as the centre of curvature of a parabolic arc over which the ball may be supposed to move for a small portion of its path, and for the corresponding velocity of rotation the governor is very sensitive. It will be seen that as the balls fly out the point at which the rods cross the axis moves upwards, as well as the balls themselves, and for a certain position of the balls the upward velocity of this point of intersection is equal to the upward velocity of the balls themselves, so that the line corresponding to AO is unchanged, and the governor therefore behaves like a parabolic governor for a small range. The student should make drawings of these different forms of governors for himself, to help him the better to understand this article.

264. *The earth's equatorial radius being taken as* 4000 *miles, it is required to find the force necessary to prevent a*

EXAMPLES OF CENTRIFUGAL FORCE. 265

particle of mass m *at the equator from leaving its surface on account of the diurnal rotation.*

The time in which the earth makes a complete rotation about its axis is a sidereal day, that is, about $\frac{365}{366}$ mean solar days, or nearly 86,164 seconds. The velocity of a point on the equator due to the rotation is therefore $\frac{2\pi \cdot 4000 \times 5280}{86164}$ feet per second. Hence the force which must act towards the centre of the earth to prevent a particle of mass m from leaving the surface must be

$$m \frac{4\pi^2 \cdot 4000^2 \cdot 5280^2}{86164^2 \cdot 4000 \cdot 5280}$$ dynamical units of force

$= \cdot 11203 m$ dynamical units of force, very nearly.

Hence the resultant force upon a particle of mass m at the equator must act towards the centre of the earth and be equal to ·11203m units of force in order that it may be at rest on the surface.

Now suppose the force with which the earth attracts the particle to be denoted by mf, and the pressure of the particle on the ground to be mg. Then the pressure of the ground on the particle is also mg, and the resultant force upon it is $mf - mg$ towards the earth's centre;

$$\therefore m(f - g) = \cdot 11203 m,$$

or the apparent weight of the particle—viz., mg—is less than the force with which the earth attracts it by ·11203m.

If we suppose the value of g at the equator to be 32, we see that the weight of a body in its neighbourhood is diminished by about $\frac{1}{286}$ of the whole weight on account of the earth's rotation, assuming the earth's radius to be 4,000 miles.

The effect of the earth's rotation upon bodies at the equator is actually to diminish their apparent weight by about $\frac{1}{293}$ of the whole. This force produces its whole effect in changing the direction of the body's motion, that is, in producing in it acceleration towards the earth's centre. The value of g at the equator is rather less than 32, but the earth's equatorial radius is only about 3962 miles instead of 4000 miles, as we have taken it.

265. Suppose a string, the mass of a unit of length of which is m, to form a circle of radius r, and to revolve in its own plane with uniform angular velocity ω. If we consider a very small length s of the string, the mass of the element will be ms, and the force which must act upon it towards the centre of the circle will be $ms\omega^2 r$. The string must therefore be acted upon by a normal force on each element towards the centre at the rate of $m\omega^2 r$ units of force per unit of length, and this force must be supplied by the *tension* of the string. The tension, T, may be determined from the following consideration. Suppose the radius of the circle to diminish by a very small quantity h. The circumference will then be reduced by $2\pi h$, and the work done by the tension will be $2\pi hT$. But if the tension be employed to balance a normal pressure of $m\omega^2 r$ units of force per unit of length, the work done against this pressure will be $2\pi r \cdot m\omega^2 r \cdot h$. Hence

$$2\pi hT = 2\pi m\omega^2 r^2 h,$$
or $\qquad T = m\omega^2 r^2 = mv^2,$

if v denote the linear velocity of the string. This result is of very great importance. It shows that in the case of a string revolving in a circle the tension depends on the linear density and the square of the velocity, and is independent of the size of the circle. This tension determines a limit to the peripheral speed of fly-wheels and the wheels of locomotives and railway carriages. If the tension in iron or steel be restricted to 4,000 lbs. per square inch, the velocity cannot exceed 197 feet per second, and no advantage is gained by increasing the diameter of the wheels.

Since the tension is unaffected by the radius of the circle, it follows that if a rope or belt be guided to move in any curve of varying curvature, the least tension consistent with the motion will still be given by the expression mv^2. In the case of a belt running over pulleys, a constant tension mv^2 will be required throughout the belt to keep it in its curved path without producing any pressure on the pulleys, and without transmitting any power. The requisite tension to produce the necessary grip of the pulleys and to transmit the power must therefore be added to this, and the result is to limit the speed of a belt of given strength and weight, and therefore the power which can be transmitted by it. For example, if the working tension of a leather belt be restricted to 400 lbs. on the square inch (about one eighth of its breaking load), the density of the leather being 64 lbs. per cubic foot, the maximum speed at which the belt can be run without transmitting any power is 170 feet per second, and the maximum power is transmitted at a speed of about 98 feet per second when the power transmitted amounts to about 32 H.P. per square inch of section of the belt.

Another very important consequence follows from the same expression for the tension in a running belt or string. If the belt is running over any guides it will hang between the guides in certain curves, and as the belt runs, these curves will remain as stationary waves upon the belt which will run past them. Now the passage of the material of the belt through the stationary wave is precisely the same action as the transmission of a wave at the same speed along a stationary belt. Hence we conclude that in a string whose linear density is m, while its tension is T, the velocity of transmission of a wave will be given by the equation

$$T = mv^2,$$
$$\text{or} \quad v = \sqrt{\frac{T}{m}}.$$

This determines the velocity of transmission of a wave in a stretched string such as the strings of a musical instrument, and from it follows at once the number of vibrations which such a string of known length can execute per second, or the pitch of the fundamental note of the string.

HARMONIC MOTION.

265a. The motion corresponding to a simple rectilinear harmonic vibration is best described as follows:—

Suppose a point P to move with uniform velocity in the circumference of a circle of which AB is a fixed diameter, and let a perpendicular PN be always let fall from P upon AB. Then as P revolves in the circle, N will execute harmonic vibrations in the straight line AB. The period of the vibration will be the same as the period of revolution of P. The *amplitude* of the oscillation is the length of AB.

Suppose that P describes the circle under the action of a force always tending towards the centre O and proportional to the distance of P from the centre, so that the acceleration of P towards the centre may be represented by $\mu . OP$. Then the acceleration of P parallel to AB will be $\mu . ON$, and its acceleration perpendicular to AB will be $\mu . PN$. But the acceleration of P parallel to AB is the same as the acceleration of N along AB. Hence N moves with an acceleration $\mu . ON$ towards O, and always proportional to its distance from O.

Again, if v denote the velocity of P, its acceleration towards O must be $\dfrac{v^2}{OP}$. Hence

$$\frac{v^2}{OP} = \mu OP$$

or $v = \sqrt{\mu} . OP$.

Hence the time occupied by P in making a complete revolution is $\dfrac{2\pi}{\sqrt{\mu}}$, and is independent of the dimensions of the circle. The time occupied by N in making a complete oscillation under the action of a force producing an acceleration towards a fixed point represented by μ times the distance of N from that point is therefore $\dfrac{2\pi}{\sqrt{\mu}}$, and is independent of the extent, or amplitude, of the oscillation.

Again, the velocity of N is to that of P as PN to OP. The velocity of N is therefore equal to $\sqrt{\mu} . OP \dfrac{ON}{OP}$, that

HARMONIC MOTION. 269

is, to $\sqrt{\mu}\, ON$, and is proportional to $\sin PON$. The acceleration is proportional to ON, that is, to $\cos PON$, and the displacement from O is also proportional to $\cos PON$.

Rectilinear harmonic motion may be regarded as the projection of uniform circular motion on a plane perpendicular to the plane of the circle.

The above investigation of harmonic motion affords a proof, quite independent of the properties of the cycloid, of the theorem that if a point move with an acceleration towards a fixed point and always proportional to its distance from that point, the time of oscillation is independent of the amplitude.

266. When two or more bodies are so connected that if the motion of one of them be given that of each of the others is known, we can, by help of the equations expressing the geometrical connections of the system, find the motion of each part and the forces between the parts when the external forces acting on the system are known. We have seen examples of this in the cases of weights connected by a string over a pulley. The general method of solution of problems of this class is to take the acceleration of one of the parts as the unknown quantity, then by help of the geometrical equations the accelerations of all the other parts can be expressed in terms of this. The accelerations of all the parts being thus expressed in terms of one unknown quantity, the *resultant force* upon each can also be so expressed; hence the reactions between the parts can be expressed in terms of this one unknown; and finally, the resultant force on the first part of the system being expressed in terms of the external forces upon it, and the reactions of the other parts of the system, it can be expressed in terms of the unknown acceleration. But it is this force which produces that acceleration in the first part of the system, and this furnishes us with another expression for the same force. Equating these two expressions we have an equation to determine the unknown acceleration. This process will be best understood by an example.

267. Ex. *A smooth wedge A whose angle is a and mass*

M *rests on a smooth plane inclined at the same angle a to the horizon, so that one face of the wedge is horizontal. On the upper surface of the wedge is placed a weight* B *of mass* m. *Find the motion of the system and the pressures between the parts.*

The upper surface of the wedge being smooth, all the forces upon the weight B are vertical, and therefore this weight will descend in a vertical straight line. Also, since the weight remains on the top of the wedge, which is horizontal, the vertical motion of the wedge must be the same as that of the particle, and therefore its vertical acceleration must be the same. Also the motion of the wedge is always along the inclined plane, and therefore its acceleration must be in that direction.

Let f denote the acceleration of B, f' that of the wedge A, then, since their vertical accelerations are the same,

$$f' \sin a = f;$$

therefore

$$f' = \frac{f}{\sin a}.$$

Therefore the resultant force on the wedge must be $\dfrac{Mf}{\sin a}$ along the plane. Let P denote the pressure of the weight B on the wedge. Then resolving along the plane,

$$\frac{Mf}{\sin a} = (P + Mg) \sin a;$$

$$\therefore P = \frac{Mf}{\sin^2 a} - Mg.$$

But the resultant force upon B is $mg - P$, acting vertically downwards, and this must therefore be the force required to produce an acceleration f in the mass m. Hence

$$mg - \frac{Mf}{\sin^2 a} + Mg = mf,$$

or

$$f = \frac{(M + m) g \sin^2 a}{M + m \sin^2 a};$$

this determines the acceleration of the weight B. The acceleration of the wedge along the plane is

$$\frac{(M+m)g\sin a}{M+m\sin^2 a},$$

and the pressure, P, between the weight B and the wedge is
$$\frac{mMg\cos^2 a}{M+m\sin^2 a}.$$

Resolving perpendicularly to the inclined plane we see that the pressure between the wedge and plane is

$$(P+Mg)\cos a,$$

that is, $$M\frac{(m+M)}{M+m\sin^2 a}g\cos a.$$

268. When the connections between the parts of a system are such that the motion of all can be expressed in terms of that of one of them, and there are no sudden changes of velocity in any part of the system, we may frequently determine the motion from the consideration that the kinetic energy of the system is equivalent to the work done upon it by external forces. We may illustrate this by the following example.

269. Ex. *A weight of 64 lbs. is supported in equilibrium by a weight of 4 lbs. in a system of pulleys in which each string is vertical. If a half-pound weight be added to the 4 lb. weight, determine the motion of the system, neglecting the friction and inertia of the pulleys, strings, etc.*

Since a weight of 4 lbs. supports in equilibrium a weight of 64 lbs. it follows from the principle of vertical velocities that if the former fall through a very small space the latter will rise through $\frac{1}{16}$ of that space, and since the strings are vertical the system is always similar throughout the motion. Hence the geometrical connections must be such that in each displacement throughout the motion the 4 lb. weight will move through 16 times the space moved through by the 64 lb. weight, and its velocity and acceleration will therefore be 16 times that of the larger weight. Hence, since the geometrical connections are undisturbed by adding the half-pound weight, the same will be true in the motion we are considering. Let f be the acceleration of the $4\frac{1}{2}$ lbs. Then $\dfrac{f}{16}$ is the acceleration of the 64 lb. weight, and f will remain constant throughout the motion, since the conditions are always the same. Hence the kinetic energy of the system at the end of time t from the commencement of the motion will be

$\dfrac{1}{2} \cdot 4\frac{1}{2} \cdot f^2 t^2 + \dfrac{1}{2} \cdot 64 \cdot \dfrac{f^2 t^2}{16^2}$, or $\dfrac{19}{8} f^2 t^2$, units of kinetic energy.

Also, the space described by the $4\frac{1}{2}$ lb. weight will be $\dfrac{1}{2} f t^2$ units, and that through which the 64 lb. weight has ascended will be $\dfrac{1}{2} \cdot \dfrac{f}{16} t^2$. Hence the work done on the system by gravity will be $\dfrac{1}{2} g \dfrac{1}{2} f t^2$ units of work.

Therefore $\dfrac{1}{4} g f t^2 = \dfrac{19}{8} f^2 t^2$; $\therefore f = \dfrac{2}{19} g$,

and the $4\frac{1}{2}$ lbs. will descend with uniform acceleration $\dfrac{2}{19} g$.

270. Or we may proceed thus:—The tension of the string supporting the "weight" will always be 16 times that of the string supporting the "power," whether the system be at rest or in motion, since the weights of the strings and pulleys are neglected. Let T denote the latter tension in absolute units, then $16T$ will represent the former. The acceleration of the "power" will therefore be $g - \dfrac{T}{4\frac{1}{2}}$ downwards, and that of the "weight" $\dfrac{16T}{64} - g$ upwards. But the acceleration of the "power" is always 16 times that of the weight;

$\therefore g - \dfrac{T}{4\frac{1}{2}} = 16\left(\dfrac{16T}{64} - g\right)$, or $T = \dfrac{17 \times 9}{38} g$,

and the acceleration of the $4\frac{1}{2}$ lb. weight is $g - \dfrac{T}{4\frac{1}{2}}$, or $\dfrac{2}{19} g$, as before.

INITIAL ACTIONS. 273

271. We purpose now to give a few examples of a class of problems not unfrequently proposed, namely the following :—

Suppose a system of particles at rest and in equilibrium under given constraints, and let one of these constraints be suddenly removed. It is required to find the change instantaneously produced in the action of the other constraints.

The general method to be adopted in order to determine the initial actions of the remaining constraints is to find the direction and acceleration with which each particle *begins* to move. If we multiply the expression for this acceleration by the mass of the particle, the product is the measure of the resultant force upon the particle, and this resultant force being determined in magnitude and direction, we have sufficient equations for determining all the forces in the system at the *commencement* of the motion.

272. Ex. 1. *A particle of mass* m *is suspended from two points in the same horizontal line by two strings of equal lengths. One of the strings is suddenly cut. It is required to find the initial change of tension of the other string.*

Let P be the particle in its position of rest, A, B the

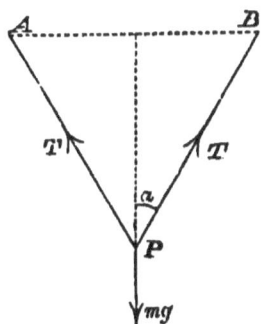

points of suspension. Let $AB = 2a$, and let the length of each string be l. Let the angle BPA be equal to $2a$. If

G. D. T

T be the tension of each string when there is equilibrium, we have by resolving vertically

$$T = \frac{mg}{2\cos a},$$

where $$\cos a = \frac{\sqrt{l^2 - a^2}}{l}.$$

Now suppose the string AP suddenly cut. Then BP remaining of invariable length, the particle will begin to move at right angles to BP. Hence the resultant force upon it must be in this direction. Therefore, if T' be the tension of the string BP, immediately after cutting AP, we have

$$T' = mg \cos a,$$

and the initial change of tension is

$$T' - T = mg \left(\frac{2\cos^2 a - 1}{2\cos a} \right).$$

If a be less than $45°$ the tension of BP is suddenly increased, and if a be greater than $45°$, or l less than $\sqrt{2}.a$, the tension of BP is suddenly diminished, by cutting AP.

The resultant force upon the particle immediately after cutting the string AP is $mg \sin a$, acting in a direction perpendicular to BP. Hence the initial acceleration of the particle is in this direction and is numerically equal to $g \sin a$.

273. Ex. 2. *A string having its ends fastened to two fixed points* A *and* B *in the same horizontal straight line has four equal particles, each of mass* m, *attached to it at equal intervals. If while the system is at rest the string be cut in the middle, it is required to find the instantaneous change of tension of the other portions of the string.*

When at rest the portion QR of the string will be horizontal, and the system will be symmetrical about the vertical line through the middle point of the string. Let a_1 denote the inclination of RS to the horizon, a_2 that of BS. Then, if T_0 represent the tension of QR when the

INITIAL TENSIONS. 275

system is in equilibrium, T_1 that of RS, and T_2 that of BS, we have

$$T_2 \cos a_2 = T_1 \cos a_1 = T_0 \ldots\ldots\ldots\ldots(1).$$

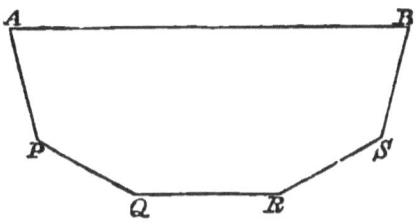

Also $\quad T_1 \sin a_1 = mg$, or $T_1 = \dfrac{mg}{\sin a_1}$,

and $\quad T_2 \sin a_2 - T_1 \sin a_1 = mg$,

or $\quad T_2 = \dfrac{2mg}{\sin a_2}.$

If the length of the string be known as well as the distance AB, we have sufficient equations for determining a_1 and a_2.

Let T_1' represent the tension of RS, and T_2' that of SB, immediately after cutting the string between Q and R. Then since SB remains of invariable length, the direction in which S moves is always perpendicular to SB, and hence *at the beginning of the motion* its acceleration along SB must be zero.

Therefore $\quad T_2' - T_1' \cos(a_2 - a_1) = mg \sin a_2$;

therefore $\quad T_2' = mg \sin a_2 + T_1' \cos(a_2 - a_1).$

Also the resultant force upon S perpendicular to BS is

$$mg \cos a_2 - T_1' \sin(a_2 - a_1),$$

and its acceleration in this direction is therefore

$$g \cos a_2 - \dfrac{T_1'}{m} \sin(a_2 - a_1),$$

and the component of this in the direction RS is

$$\left\{ g \cos a_2 - \frac{T_1'}{m} \sin (a_2 - a_1) \right\} \sin (a_2 - a_1).$$

Now since SR is of invariable length the velocity of R in the direction RS must always be the same as that of S in that direction. Hence the acceleration of R in the direction RS must be equal to that of S in the same direction. The resultant force upon R in the direction RS is

$$T_1' - mg \sin a_1,$$

and its acceleration in this direction is therefore

$$\frac{T_1'}{m} - g \sin a_1.$$

Hence we have

$$\frac{T_1'}{m} - g \sin a_1 = g \cos a_2 \sin (a_2 - a_1) - \frac{T_1'}{m} \sin^2 (a_2 - a_1);$$

therefore

$$T_1' = mg \frac{\sin a_1 + \cos a_2 \sin (a_2 - a_1)}{1 + \sin^2 (a_2 - a_1)}$$

$$= mg \frac{\sin a_2 \cos (a_2 - a_1)}{1 + \sin^2 (a_2 - a_1)}.$$

But $\quad T_2' = mg \sin a_2 + T_1' \cos (a_2 - a_1);$
therefore

$$T_2' = mg \frac{2 \sin a_2}{1 + \sin^2 (a_2 - a_1)}.$$

Hence we have found the tensions of the parts of the string immediately before and immediately after the section. The differences between the corresponding tensions will of course be the instantaneous changes required.

By proceeding precisely in the same way we might find the tension of the different portions of the string, immediately after cutting it at *any* point, whatever be the number of weights suspended from it.

PRESSURE EXERTED BY A STEAM HAMMER. 277

274. We will conclude this chapter with the following example, illustrating the application of the theory of energy to problems connected with uniformly accelerated motion.

Ex. *A Nasmyth hammer, moving in a vertical direction, is driven by steam pressure on a circular piston 40 inches in diameter. The mass of the hammer and piston together is 25 tons, and the pressure of steam on the upper side of the piston is equal to the weight of 50 lbs. per square inch more than on the lower side. Supposing the hammer after falling through 59 inches to strike a mass of iron, compress it vertically through one inch, and then come to rest, find the mean pressure exerted by the hammer upon the iron.*

The whole force acting upon the hammer before striking the iron is equal to the weight of

$$25 + \frac{\pi \cdot 20^2 \cdot 50}{2240} \text{ tons,}$$

and the whole mass moved being 25 tons, it will fall with uniform acceleration represented by

$$g\left(1 + \frac{\pi \cdot 20^2 \cdot 2}{2240}\right) \text{ or } g\left(1 + \frac{5\pi}{14}\right).$$

The velocity, v, which the hammer will acquire in falling through 59 inches, is given by the equation

$$v^2 = 2g\left(1 + \frac{5\pi}{14}\right)\frac{59}{12},$$

and this velocity is destroyed by a constant force, while the hammer moves over one inch. Hence if f denote the uniform acceleration with which the hammer moves throughout that inch, we have

$$v^2 = 2f \cdot \frac{1}{12};$$

$$\therefore \frac{1}{6}f = 2g\left(1 + \frac{5\pi}{14}\right)\frac{59}{12},$$

$$\text{or } f = 59g\left(1 + \frac{5\pi}{14}\right).$$

278   PRESSURE EXERTED BY A STEAM HAMMER.

Therefore the resultant force on the hammer, while compressing the iron, is equal to the weight of

$$25 \times 59 \left(1 + \frac{5\pi}{14}\right) \text{ tons.}$$

But the downward force upon it due to its weight, and the pressure of the steam, is equal to the weight of

$$25 \left(1 + \frac{5\pi}{14}\right) \text{ tons.}$$

Hence the vertical pressure of the iron on the hammer, and therefore of the hammer on the iron, is equal to the weight of $60 \times 25 \left(1 + \frac{5\pi}{14}\right)$ tons, or about 3183 tons.

275. We might have obtained the result of the preceding article from the consideration that if a mass move under uniform acceleration, the change of its kinetic energy in any time is always numerically equal to the work done upon it during the interval. If the acceleration under which the body is moving be suddenly changed during the motion, this principle is true for each portion of the motion, and therefore throughout the whole.

Now the hammer starts from rest, and finally comes to rest. Hence the whole work done upon it must be zero. But it falls altogether through 5 feet under a constant force equal to the weight of $25\left(1 + \frac{5\pi}{14}\right)$ tons. Hence the work done upon it by this downward force is $125 \left(1 + \frac{5\pi}{14}\right)$ foot-tons. Therefore the work which the hammer must do upon the iron is $125 \left(1 + \frac{5\pi}{14}\right)$ foot-tons. But it compresses the iron through $\frac{1}{12}$ of a foot. Hence the mean pressure which it exerts upon the iron must be equal to the weight of

$$1500 \left(1 + \frac{5\pi}{14}\right) \text{ tons.}$$

EXAMINATION ON CHAPTER V.

1. A particle is projected from the vertex of a smooth parabolic tube, whose axis is vertical, and latus rectum equal to $4a$, along the tube with a velocity represented by $\sqrt{2ag}$. Find the velocity of the particle at any point in the tube in terms of the focal distance of the point.

2. Assuming that on descending a mine, g varies directly as the distance from the earth's centre, find the number of beats lost in a day by a pendulum which beats seconds at the sea-level, when carried down a mine to a depth of 400 fathoms, supposing the earth a sphere of 4000 miles radius.

3. Show that the time of oscillation of a particle under the action of gravity about the lowest point of a vertical circle of radius $2a$, is greater than the time of oscillation on a cycloid the diameter of whose generating circle is a, if the arc of oscillation in the circle be of finite length.

4. A particle slides down the surface of a right circular cylinder whose axis is horizontal from rest on the highest generating line. Find the pressure on the cylinder in any subsequent position of the particle, and the point where the particle will leave the surface.

5. Supposing the mass of the bob of a conical pendulum to be 20 lbs., and the length of the string to be 3 feet, find the inclination of the string to the vertical when the bob is making 3 revolutions per second, and its tension.

6. A heavy particle is attached to a string 5 feet long, and swung round in a vertical circle. Find its velocity at the highest point in order that the string may just remain tight.

7. Explain the action of the conical pendulum as a regulator or "governor" for a steam engine.

8. In the case proposed in question 1, find the pressure of the particle upon the tube at any point of its path.

280   EXAMPLES ON CHAPTER V.

9. A train goes round a curve whose radius (*i.e.* the radius of the curve lying midway between the two metals) is 150 yards, at the rate of 50 miles per hour. Find the height to which one of the metals must be raised above the other in order that the whole pressure of each carriage on the metals may be perpendicular to the floor of the carriage, the breadth of the gauge being 4 ft. $8\frac{1}{2}$ ins.

10. A heavy particle is suspended from the angular points of an equilateral triangle whose plane is horizontal, by means of three strings each equal in length to one side of the triangle. If one of the strings be cut, find the initial change of tension of the other two.

11. A uniform endless string, of length $2\pi a$, is rotating in its own plane with uniform angular velocity ω, under the action of no external forces. Find the tension of the string, the mass of each unit of length being m.

12. The mass of a smooth wedge whose angle is 30° is 10 lbs., and it rests on a smooth plane inclined 30° to the horizon, so that the upper surface of the wedge is horizontal. A weight of 2 lbs. is placed on the top of the wedge. Find its acceleration and the pressure of the weight on the wedge.

EXAMPLES ON CHAPTER V.

1. The value of g at Greenwich being 32·1912 and at Trinidad 32·0913, find how many beats a Greenwich seconds' pendulum would lose in a day at Trinidad.

2. Show that the acceleration of a particle oscillating in a smooth cycloidal tube whose axis is vertical, is at any point proportional to its distance from the vertex measured along the curve.

3. Find the inclination to the vertical of a conical pendulum 20 inches long, and making 200 revolutions per minute.

4. A body whose mass is 10 lbs. is suspended by a string from a point in the roof of a railway carriage, which is describing a curve of 509 feet radius at the rate of 45

miles an hour. Find the inclination of the string to the vertical when it is in relative equilibrium, and the tension of the string.

5. Find the difference in the pressures exerted on the metals by a train of 200 tons when going due East, and when going due West, along a horizontal rail at 60 miles an hour in latitude 60°.

6. A pendulum which at A beats seconds, gains 2 beats an hour at B. Compare the weights of the same substance at the different places.

7. Two very small imperfectly elastic balls are let fall simultaneously from different points, their centres moving on the same cycloid whose axis is vertical and vertex downwards. Show that all their impacts will take place at the vertex, and find the ultimate range of vibration when the impacts have ceased.

8. The length of a pendulum which vibrates 30 times in a minute in 156·8 inches. Find the space through which a particle will fall from rest in one second under the action of gravity.

9. A free body falls from rest through nearly $301\frac{1}{4}$ yards in one-eighth of a minute in the latitude of Greenwich. How far would a body fall from rest in a quarter of a minute at a place where the length of the seconds' pendulum is ·999 of its length at Greenwich?

10. The attraction of a planet of mass m on a given body at a point distant r from its centre, r being greater than the radius of the planet, varies as $\dfrac{m}{r^2}$. The mass of the earth is 49 times that of a certain planet, while its radius is 4 times that of the planet. Prove that a seconds' pendulum carried to the planet would oscillate in about $\dfrac{7}{4}$ seconds.

11. If a simple pendulum $39\frac{1}{3}$ inches long oscillate in one second, what is the length of a pendulum which makes 3540 beats in an hour?

12. The horizontal attraction of a mountain on a par-

ticle at a certain place is such as would produce in it an acceleration denoted by $\frac{1}{n}g$. Show that a seconds' pendulum at that place will gain $\frac{21600}{n^2}$ beats in a day, very nearly.

13. Show that a pendulum one mile long would oscillate in about $\frac{121}{120} \cdot \frac{1}{7} \sqrt{22}$ minutes.

14. A seconds' pendulum is carried to the top of a mountain 3000 feet high; assuming that the force of gravity varies inversely as the square of the distance from the earth's centre, and that the earth's radius is 4000 miles, find the number of oscillations lost in a day, neglecting the attraction of the mountain.

15. A railway train is moving uniformly along a curve at the rate of 60 miles per hour, and in one of the carriages a pendulum which would ordinarily beat seconds, is observed to oscillate 121 times in two minutes. Show that the radius of the curve is very nearly a quarter of a mile.

Supposing a stone dropped from the window of one of the carriages, find how much farther from the centre of the curve is the point at which it strikes the ground than the point vertically beneath that from which it falls, the height of the latter point above the ground being 6 feet.

16. A particle is projected horizontally with a given velocity from the highest point of a smooth sphere. Find the point where it leaves the sphere.

17. Find the greatest velocity with which a particle may be projected horizontally from the highest point of a sphere, so as to begin to move on the surface of the sphere.

18. A smooth straight tube is made to describe a right circular cone whose axis is vertical and semivertical angle equal to a, with uniform velocity, the vertical plane through the tube turning about the axis of the cone with uniform velocity ω. Find where a particle will be in relative equilibrium in the tube.

EXAMPLES. 283

19. Show that if a heavy particle fall from a cusp down the arc of a smooth cycloid whose axis is vertical and vertex downwards, its pressure on the curve at its lowest point will be equal to twice its weight.

20. Supposing the earth's orbit about the sun to be a circle of 93,000,000 miles radius, and the earth to describe this orbit with uniform velocity in $365\frac{1}{4}$ days, express the force exerted by the sun on a pound of matter at the earth's surface in British absolute units, neglecting the magnitude of the earth in comparison with the sun's distance.

21. If different points be describing different circles uniformly with accelerations proportional to their radii, their periodic times will be the same.

22. A heavy particle is placed very near the highest point of a smooth vertical circle; show that the latus rectum of the parabola which it describes after leaving the circle is to the radius of the circle as $16 : 27$.

If, retaining the same highest point, the circle vary in size, show that the locus of the focus of the parabolic path of a particle so flying off is a straight line.

23. A lamina in the form of a regular hexagon of side a is placed flat on a smooth horizontal plane and fastened to the plane. A string of length equal to the perimeter of the polygon is wound round it, one end being attached to an angular point, and the other end carrying a particle of mass m. If the particle be projected horizontally at right angles to the string with velocity v, find the time after which the string will be wound up again, and its greatest and least tensions.

24. A skater, whose weight is 12 stone, cuts on the outside edge a circle of 3 yards radius, with uniformly decreasing velocity, just coming to rest after completing the circle in 6 seconds. Find the direction and magnitude, when he is half-way round, of his pressure on the ice.

25. A particle suspended from a point by a string of length a is projected from its lowest position with velocity $\dfrac{\sqrt{3}+1}{\sqrt{2}}\sqrt{ga}$; show that it will pass through the point of

suspension, and that the direction of its motion at that point will make an angle $\cos^{-1}\frac{1}{3}$ with the horizon.

26. If a wheel of radius a roll on the lower side of a horizontal plane so that its centre moves in a straight line with uniform velocity $\sqrt{ga}$, any point on its circumference will move in the same manner as a heavy particle starting from the cusp of a smooth cycloidal arc whose axis is vertical, and sliding down it.

27. A particle starts from the extremity of a smooth cycloidal arc whose axis is vertical; show that when it has fallen through half the distance measured along the arc to the vertex, it will have accomplished ¾ of its vertical descent, and two-thirds of the time of descent will have elapsed.

28. Three equal smooth spheres are placed in contact on a horizontal plane, and are connected where they touch. A fourth equal smooth sphere is placed so as to be supported by the other three. If the connections between the lower spheres be simultaneously broken, show that the pressure between each and the upper sphere is instantaneously diminished by one-seventh.

29. A heavy uniform string rests on a smooth horizontal table with one end pinned to the table and $\frac{1}{n}$ of its length hanging over the edge of the table; if the pin be removed the resultant pressure on the table will be instantaneously diminished by $\frac{1}{n^2}$ of the weight of the string.

30. A smooth wedge, whose vertical angle is 30° and mass 10 lbs., is placed on a smooth plane inclined 45° to the horizon, the edge of the wedge being horizontal and directed upwards. On the top of the wedge (which is inclined at an angle of 15° to the horizon) is placed a smooth weight of 5 lbs. Determine the motion, and the pressures between the weight and the wedge, and between the wedge and the plane.

31. In a system of pulleys in which all the strings are

vertical, a weight of one pound can support a weight of 32 lbs. If a weight of one ounce be added to the one-pound weight, determine the motion of the system and the space through which the 32 lb. weight will be raised in one minute, neglecting the friction and inertia of the pulleys, and the rigidity and inertia of the ropes.

32. A small smooth ball is running horizontally round the inside of a hemispherical bowl of given size. Show how to determine its height above the bottom of the bowl by observation of the time taken in making each circuit.

33. A circular elastic band is placed round a wheel, the circumference of which is twice the natural length of the band; if the wheel be made to revolve with constant angular velocity, find the pressure of the band on the wheel.

34. A heavy particle is placed very near the vertex of a smooth cycloid having its axis vertical and vertex upwards; find where the particle runs off the curve and prove that it falls on the base of the cycloid at the distance $\left(\frac{\pi}{2} + \sqrt{3}\right) a$ from the centre of the base, a being the radius of the generating circle.

35. Pendulums which beat seconds correctly in London ($g = 32{\cdot}19$) and Edinburgh ($g = 32{\cdot}20$) respectively are interchanged in station. If started simultaneously from the vertical position towards the left, after how many seconds will they again be both vertical and moving leftwards?

36. A simple seconds'-pendulum is formed by a particle of given mass suspended by a string. If the length of the string in any latitude be the unit of length and its tension the unit of force, and if the unit of density be constant, prove that the unit of energy is inversely as the unit of acceleration.

37. A parabola is placed with its plane vertical, its vertex upwards and its axis inclined at an angle a to the horizon: if a particle start from rest at the highest point it will leave the curve when its direction of motion makes an angle equal to $\tan^{-1} \sqrt[3]{\tan a}$ with the horizon.

CHAPTER VI.

APPENDIX OF THE DYNAMICAL THEORY OF GASES.

276. A SIMPLE gas may be considered to be a collection of equal, free, material particles moving about in all directions. When two of these particles approach one another they behave like perfectly elastic balls impinging on each other, that is to say, the velocity of either relative to the other becomes reversed in direction but is unaltered in magnitude. We do not say that the particles *are* perfectly elastic, because we cannot conceive of a *particle* suffering compression and subsequently regaining its original form. It is probable that when the particles approach very near one another a repulsive force acts between the two, which increases as the distance between the particles diminishes, so as to prevent their ever coming

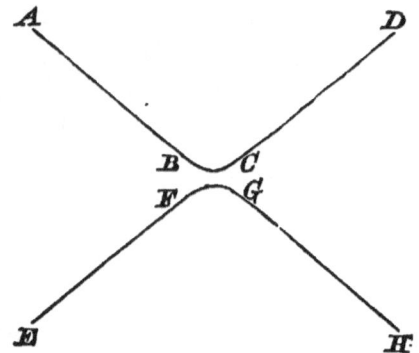

into actual contact, and this force, depending only on the distance between the particles, and not on the direction of their motion, will generate as much momentum in each as they recede from one another as it destroyed during their approach. If the particles are not moving in the same straight line, since the force between them does not

DENSITY OF A GAS. 287

act in the direction of the motion of either, the path of each will be curved throughout that portion during the description of which the force is sensible. Thus the paths of two particles in the neighbourhood of the least distance between them will be represented by the lines $ABCD$, $EFGH$, where AB, EF are the directions of motion before coming sufficiently near together to act sensibly on one another, and CD, GH the directions in which they are moving after getting beyond the reach of each other's repulsion.

When the particles of a gas meet with any obstacle, whose mass is very great compared with their own, they behave like elastic balls projected against a fixed surface, and rebound, their velocity relative to the obstacle remaining unaltered in magnitude but reversed in direction.

277. If a unit of volume of a gas contain n particles, each of mass m, then the mass of the unit of volume will be nm. Now it may be shown that if a quantity of gas of mass M be contained within a closed vessel, the resultant pressure upon the vessel is Mg, that is, the weight of the gas. Hence this can be determined by weighing, and then M will become known. Hence if the volume of the vessel be known the mass of unit volume can be found. Thus nm can be found for any particular gas under given circumstances (i.e. given pressure and temperature).

All we know about n is that it lies between certain limits, which are however so very far apart as to be of little value. It may be assumed that, if the unit of volume be a cubic foot, n is many trillions at least. Since we are unable to determine n, we are equally unable to determine m. Hence all questions relating to the pressure of gases upon surfaces, and the like, must be treated in the same manner as the example in Art. 228.

A cubic millimetre of gas at the ordinary temperature and pressure probably contains about 5×10^{18} molecules.

In fact, as we are unable to isolate particular particles (or molecules as they are called) of a gas and trace out their personal history, the collisions with other particles which they experience, and the like, we have to adopt a statistical method, and fixing our attention upon some particular condition, we determine what proportion of the whole number of particles under our consideration fulfils that condition. Thus we may consider what proportion of the whole number of particles in a unit of volume impinges upon a given surface during a second, or what proportion impinges upon a given surface the angles of incidence upon it lying between given limits.

The quantity of matter in the unit of volume of a substance is the measure of its density. Hence we may call nm the density of the gas, and we shall use this term for nm in the following articles, and denote it by ρ.

278. Before considering the pressure exerted by a gas on any area, the particles of the gas moving in directions uniformly distributed throughout space, we shall consider the pressure exerted on the unit of area by an imaginary gas of density ρ, all the particles of which are running in a direction perpendicular to the surface and with the same velocity.

Each cubic foot of the gas contains a mass represented by ρ or nm, a foot being the unit of length. Now of the n particles in the unit of volume one half are moving towards the plane and one half away from the plane. Let v be the velocity of each particle in feet per second. Then since each cubic foot of gas contains $\frac{n}{2}$ particles moving towards the plane with velocity v, the number of particles which strike a square foot of the plane in a second will be $\frac{nv}{2}$ and their mass $\frac{nm}{2}v$. Now after impact they each move *from* the plane with velocity v. Hence the change of momentum of each particle produced by the action of the plane is $2mv$, and therefore the whole change of momentum produced in one second by the pressure of the plane on the gas is $\frac{nv}{2} \cdot 2mv$, that is nmv^2, and this is therefore the measure of that pressure. Hence the pressure exerted by the gas on each square foot of the plane is nmv^2, or since nm is equal to ρ, the measure of the pressure is ρv^2.

279. Next imagine an unlimited volume of gas, of density ρ, to consist of particles all of which are moving in directions making the same angle with the normal to a fixed plane, and with the same velocity. We propose to find the pressure exerted by such an imaginary gas upon each unit of area of the plane.

As in the previous case, of the nm particles contained in each cubic foot one half will be moving towards the plane, and one half away from the plane. At points very close to the plane the particles moving towards it are those just about to impinge upon it, while those moving from it are those which have just impinged. These latter, meeting with others of equal mass coming towards the plane, and each acting like a perfectly elastic ball, will exchange the components of their velocities perpendicular to the plane, and so on throughout the gas, so that while in every portion of gas one half of the particles are moving from the plane, it does not follow that these have just, or ever, impinged upon the plane. In fact each particle without moving to any sensible distance from its original position may have the direction of its motion changed any number of times.

In each cubic foot the number of particles moving towards the plane is $\frac{n}{2}$ and their aggregate mass $\frac{nm}{2}$, also the resolved part of the velocity of each perpendicular to the plane is $v \cos a$, hence the number of particles which strike each square foot of the plane in one second is $\frac{n}{2} v \cos a$, and their aggregate mass $\frac{nm}{2} v \cos a$. Now the change produced in the velocity of each by impact is denoted by $2v \cos a$. Hence the whole change of momentum produced in one second by the pressure of a square foot of the plane is $nmv^2 \cos^2 a$ or $\rho v^2 \cos^2 a$, which therefore is the measure of the pressure of the gas on the plane.

280. If an unlimited number of lines be drawn from the centre of a sphere in directions uniformly distributed throughout space, the number of such lines passing through any given area of the spherical surface will be

proportional to that area, and will therefore bear to the whole number of lines the same ratio as the given area bears to the whole area of the sphere. This must be considered the fundamental notion of uniform distribution of directions in space.

We have seen that if the directions of motion of all the particles of a gas were to make the same angle a with the normal to a plane, and the velocities of all the particles were the same, the pressure on each unit of area of the plane would be represented by $\rho v^2 \cos^2 a$. Now in all actual gases the directions of motion of the several particles are uniformly distributed in space, and in order to determine the pressure of such a gas on a plane area exposed to it, we have to determine the average value of the expression $\rho v^2 \cos^2 a$ under this hypothesis. This we proceed to do, and in the first place we shall suppose the velocities of all the particles to be the same.

281. Let O be the centre of a sphere whose radius we will suppose for simplicity numerically equal to v. Let AB be the diameter of the sphere which is perpendicular

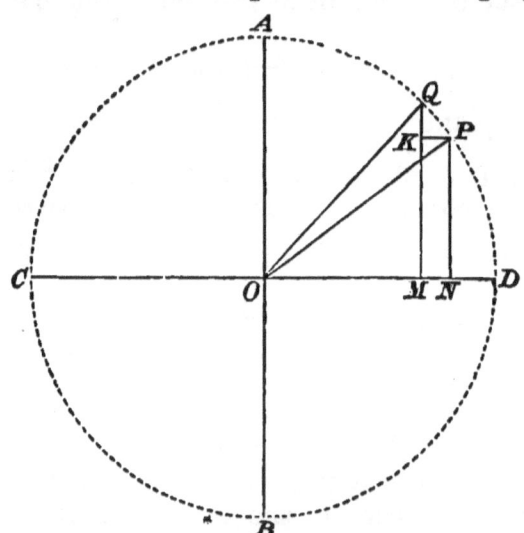

to the plane, the pressure on which we desire to measure. Let P be any point on the sphere, Q a point near to P. Let the angle AOP be denoted by a, and POQ by θ, where θ is very small. Suppose planes perpendicular to AB to pass through P and Q. They will each cut the sphere in a small circle, and comprise between them a zone which we shall call the zone PQ. Let CD be the diameter in the plane of the paper perpendicular to AB. Draw PN, QM perpendicular to CD, and PK perpendicular to QM. Then PQ being very small indeed, we may treat it as a straight line coincident with the tangent at P to the circle APD. Also the angle QPK is equal to a, and therefore $PK = PQ \cos a$. (See also Art. 293.)

Now in a cubic foot of the gas, the number of particles moving towards the plane, and the directions of whose motion make with AO angles comprised between a and $a - \theta$, is to the whole number, n, of particles in the cubic foot as the area of the zone PQ is to the whole area of the sphere. If we denote this number by N, we have

$$N = n \, \frac{\text{area of zone } PQ}{4\pi v^2}.$$

The velocity of each of these particles perpendicular to the plane, that is in the direction OA, lies between $v \cos a$ and $v \cos \overline{a - \theta}$; hence the number of particles which strike each square foot of the plane with angles of incidence lying between a and $\overline{a - \theta}$ will lie between $Nv \cos a$ and $Nv \cos \overline{a - \theta}$, and when θ is indefinitely small, we may say that this number is equal to $Nv \cos a$, that is, to

$$n \,.\, v \,.\, \frac{\text{area of zone } PQ}{4\pi v^2} \cos a.$$

Now $PQ \cos a = QK$. Hence the area of the zone PQ is to the area of the ring generated by the revolution of PK about AO as 1 is to $\cos a$. The ring generated by PK is in fact the projection of the zone PQ upon the plane through P perpendicular to AO. We shall call this ring the ring PK.

Then $\qquad Nv \cos a = nv \,.\, \dfrac{\text{area of ring } PK}{4\pi v^2}.$

Now the velocity of each of these particles perpendicular to the plane lies between $v \cos a$ and $v \cos \overline{a - \theta}$, and when θ is indefinitely small, we may suppose it to be $v \cos a$ for each. After impact this velocity is reversed.

Hence the change of momentum of each particle produced by the impact is $mv \cos a$. Therefore the whole change of momentum produced in one second by the pressure exerted by one square foot of the plane upon particles whose directions of motion make with OA angles lying between a and $\overline{a-\theta}$ is represented by

$$nv \cdot \frac{\text{area of ring } PK}{4\pi v^2} \cdot 2mv \cos a$$

$$= 2nm \cdot v \cdot \frac{\text{area of ring } PK}{4\pi v^2} \cdot v \cos a.$$

Let this be denoted by F_a. Now $v \cos a = PN$. Hence (area of ring PK) $\times v \cos a$ is equal to the volume of the cylindrical shell generated by the revolution of the rectangle PM about AO. We shall call this the shell PM, and we have

$$F_a = 2n \cdot m \cdot v \frac{\text{volume of shell } PM}{4\pi v^2}.$$

This is therefore the measure of the pressure exerted upon each square foot of the plane by particles which impinge upon it at angles of incidence between a and $\overline{a-\theta}$.

Hence the whole pressure upon a square foot of the plane is the sum of all such quantities for values of a lying between zero and $\frac{\pi}{2}$. But when their breadth is indefinitely small, the sum of the volumes of all the cylindrical shells similar to the shell PM is equal to the volume of the hemisphere, since these shells make up the hemisphere. Hence the whole pressure Π of the gas on each square foot of the plane is given by

$$\Pi = 2nmv \cdot \frac{\text{volume of hemisphere}}{4\pi v^2}$$

$$= 2nmv \frac{\frac{2}{3}\pi v^3}{4\pi v^2}$$

$$= \frac{1}{3} nmv^2$$

$$= \frac{\rho v^2}{3}.$$

282. Now m being the mass of one of the particles, and v its velocity, $\frac{1}{2} mv^2$ is its kinetic energy, and therefore $\frac{1}{2} nmv^2$, or $\frac{1}{2} \rho v^2$, is the aggregate kinetic energy of all the particles in a cubic foot of the gas. Hence we see that if the velocities of all the particles were the same, the pressure of the gas upon a unit of area of a plane exposed to its action would be numerically equal to two-thirds the kinetic energy of a unit of volume of the gas.

Now the particles of any actual gas are not only moving in different directions, but they are also moving with different velocities. Suppose a cubic foot of the gas to contain n_1 particles each moving with a velocity equal to v_1, n_2 each moving with a velocity v_2, and so on, where $n_1 + n_2 +$ etc. $= n$. Thus, the mass of each particle being m, and the kinetic energy of a cubic foot of the gas being denoted by E, we have

$$E = \frac{m}{2} \{ n_1 v_1^2 + n_2 v_2^2 + \&\text{c.} \}.$$

If we make v^2 equal to $\dfrac{n_1 v_1^2 + n_2 v_2^2 + \&\text{c.}}{n}$, we may write

$$E = \frac{nm}{2} v^2 = \frac{\rho}{2} v^2$$

The quantity v^2 is the mean of the squares of the velocities of the particles: v is sometimes called "the velocity of mean square."

The pressure produced by the gas upon any area will be the sum of the pressures produced by the systems of particles moving with the velocities v_1, v_2, &c., respectively. Now the pressure exerted on a square foot by the particles moving with velocity v_1 we have seen to be $\frac{1}{3} n_1 m v_1^2$, where n_1 is the number of such particles in each cubic foot of gas, and similar expressions hold for the pressures exerted on a square foot by each of the other systems of particles. Hence the whole pressure, P, exerted on a square foot by the gas is given by

$$P = \frac{1}{3}(n_1 m v_1^2 + n_2 m v_2^2 + \ldots \&\text{c.})$$

$$= \frac{1}{3} n m v^2$$

$$= \frac{2}{3} E,$$

and this is true however many quantities similar to v_1 we may consider. Hence the pressure exerted by any gas on each unit of area of a plane is numerically equal to two-thirds the kinetic energy of a unit of volume of the gas.

It will be observed that the energy possessed, as well as the pressure exerted, by any gas is the same as if each particle were moving with the "velocity of mean square."

283. In the preceding article we have found the pressure exerted by a gas upon each unit of area of a plane. If a curved surface be exposed to the action of a gas, the pressure upon any small portion of the surface will be the same as if it were plane, and hence if S denote the area of the whole surface, the whole pressure upon it will be

$$\frac{1}{3} \rho v^2 S.$$

The whole pressure which a gas exerts upon any unit of area of a surface exposed to its action is called the pressure of the gas. The pressure of a gas may, under certain circumstances, vary from point to point. In such case the pressure at any point is measured by the pressure which would be exerted on the unit of area if the pressure of the gas were uniform over that area, and the same as at the proposed point.

284. If the densities of two gases be different, but their pressures the same, we have ρv^2 the same for each. Hence, the pressures being the same for each, v^2 is inversely proportional to the density of the gas. Also the kinetic energy possessed by a given volume of all gases at the same pressure is the same, for the pressure is numeri-

cally equal to two-thirds of the kinetic energy possessed by a unit of volume of the gas.

If v remain constant for any particular gas, then P, its pressure, varies directly as ρ. Now if V be the volume of a given mass M of gas, $M = \rho V$. Hence $\rho \propto \dfrac{1}{V}$, and therefore $P \propto \dfrac{1}{V}$, or PV is constant.

Again, the equation
$$P = \frac{1}{3}\rho v^2$$
may be put in the form
$$P = \frac{1}{3}\frac{M}{V}v^2,$$
or $$PV = \frac{1}{3}Mv^2.$$

Therefore the product of the pressure and the volume of a given mass of gas varies directly as its kinetic energy. But (see Besant's *Hydrostatics*, p. 73) the product of the pressure and the volume of a given mass of gas increases uniformly with the temperature. Hence the kinetic energy of a given mass of gas increases uniformly with its temperature. But the kinetic energy of a mass M varies as Mv^2. Therefore v^2 increases uniformly with the temperature. Hence the value of v for any particular kind of gas depends only on its temperature. Let T denote this temperature; then if the zero of temperature be so chosen that T and v vanish together, T will always be proportional to v^2, and if V be the volume occupied by a constant mass M of the gas, we have $PV \propto T$. The temperature is then called the *absolute* temperature of the gas. But if E denote the kinetic energy of a mass M of the gas, $PV \propto E$. Therefore $E \propto T$, or the kinetic energy of a given mass of gas varies as its absolute temperature (*i.e.* The specific heat is constant).

Again, since for a given mass of gas $PV \propto T$, if P be kept constant, $V \propto T$;—the principle of the air thermometer.

285. Before proceeding further with this subject we will find the value of v for some of the principal simple gases at a given temperature. We commence with Hydrogen. The mass of a cubic foot of hydrogen, which, at the temperature of melting ice, exerts a pressure equal to the weight of 2116·4 pounds upon each square foot of surface exposed to its action, in a place where $g = 32·2$, is known from experiment to be ·005592 pounds. The pressure upon a square foot of the surface is 2116·4 × 32·2 absolute units of force. Hence $P = 2116·4 \times 32·2$. Also the mass of a cubic foot of the gas being ·005592 pounds, we have $\rho = ·005592$. Hence, since v^2 is determined from the equation

$$P = \frac{1}{3}\rho v^2,$$

we have
$$v^2 = 3 \cdot \frac{2116·4 \times 32·2}{·005592}$$

$$= 36593916;$$

therefore
$$v = 6097;$$

or the velocity of mean square for particles of hydrogen at this temperature is a velocity of 6097 feet per second.

The velocity of some of the particles may be considerably greater than this, and that of others less, but the velocity of the majority of the particles will be not very widely different from this quantity.

The pressure of any gas we have seen to be proportional to the kinetic energy of the unit of volume of the gas. Hence, if the pressures of two different gases be the same, the value of v^2 for each will vary inversely as its density. Now the density of oxygen is found by experiment to be always 16 times that of hydrogen at the same temperature and pressure. Hence the value of v for oxygen is one-fourth that for hydrogen: that is, the velocity of mean square for oxygen at the temperature of melting ice is 1524·25 feet per second.

The density of nitrogen is 14 times that of hydrogen at the same temperature and pressure. Hence, at the temperature of melting ice, the velocity of mean square for nitrogen is $\dfrac{6097}{\sqrt{14}}$ feet per second.

286. We have seen that if P represent the pressure and V the volume of a given mass M of gas, then PV increases uniformly with the temperature. Now it is found from experiment that the increase of PV for an increase of temperature of one degree centigrade is $\dfrac{1}{273}$ of its value at the temperature of melting ice, and is the same for *all* gases. Hence the zero of absolute temperature is the same for *all* gases, and is 273° centigrade below the temperature of melting ice.

From this it follows that if T_c represent the temperature of a gas in degrees centigrade reckoned from the zero of the centigrade scale, that is, the temperature of melting ice, the absolute temperature T of the gas will be equal to $273° + T_c$.

We have seen that the value of v^2 for any particular gas is proportional to the absolute temperature. Hence if the value of v for any gas at the temperature of melting ice be known, its value at any temperature T_c can be found by multiplying this value of v by $\sqrt{273 + T_c}$.

287. Suppose a mass M of gas to be contained in a cylinder of transverse section A, closed by a piston which is made to move with a small uniform velocity u away from the gas. Suppose the piston to be at the distance x from the bottom of the cylinder.

When the piston is at a distance x from the bottom of the cylinder, the volume occupied by the gas is Ax, and its density ρ is therefore equal to $\dfrac{M}{Ax}$.

Suppose a cubic foot of the gas to contain n_1 particles moving with velocity v_1, in directions making an angle a_1 with the axis of the cylinder. Then of these particles $\dfrac{1}{2}n_1$ are moving towards the piston. The velocity of each of these relative to the piston is $v_1 \cos a_1 - u$. Hence the number of these particles which strike each square foot of the area of the piston in a second is $\frac{1}{2} n_1 (v_1 \cos a_1 - u)$.

Now by the impact the velocity of each relative to the piston is reversed. The change of the velocity of each is therefore $2(v_1 \cos a_1 - u)$; and if m be the mass of each the pressure exerted on each unit of area of the piston by particles moving with velocity v_1 in directions making an angle a_1 with the axis of the cylinder will be $n_1 m (v_1 \cos a_1 - u)^2$. Now during any small time τ the piston will move over a space $u\tau$ in the direction of this pressure, and u and τ being both small we may suppose the pressure to remain uniform during this displacement. Hence the work done during the time τ upon the piston by the pressure exerted by the particular set of particles under our consideration is $A \cdot n_1 m (v_1 \cos a_1 - u)^2 \cdot u\tau$.

Again, the component of the velocity of each particle perpendicular to the axis of the cylinder is unaltered by impact on the piston while the component parallel to this axis is changed from $v_1 \cos a_1$ to $v_1 \cos a_1 - 2u$. Hence the kinetic energy lost by each particle on account of the collision is

$$\frac{m}{2} \{(v_1 \cos a_1)^2 - (v_1 \cos a_1 - 2u)^2\} = 2mu(v_1 \cos a_1 - u).$$

Also, the rate at which these particles strike the piston is $\frac{n_1}{2}(v_1 \cos a_1 - u)$ per second on each unit of area; and therefore the number which strike the piston in the small time τ is $\frac{n_1}{2}(v_1 \cos a_1 - u) A\tau$; and since the kinetic energy lost by each particle is $2mu(v_1 \cos a_1 - u)$, the kinetic energy lost by the system of particles we are considering is numerically equal to $n_1 m (v_1 \cos a_1 - u)^2 Au\tau$, that is, to the work done upon the piston by this particular set of particles.

Similarly, if we consider the set of particles moving with any other velocity, and whose directions of motion

EXPANSION ACCOMPANIED BY COOLING. 299

make any other angle with the axis of the cylinder, we obtain the same result; and, this being true for each set of particles, is true for the whole gas. Hence the loss of kinetic energy of the gas within the cylinder during any small time τ is equal to the work done upon the piston by the pressure.

Also, this being true for each small interval of time τ, it follows that the whole work done upon the piston during any finite time is numerically equal to the whole loss of kinetic energy sustained by the gas.

If the gas were contained within any other form of envelope, and this were allowed to expand in any way, it might be shown that the whole work done upon the envelope by the pressure of the gas is numerically equal to the kinetic energy lost by the gas.

288. Since the temperature of a given mass of gas is proportional to its kinetic energy, it follows that if a gas be allowed to expand and do work upon the vessel which contains it, the temperature of the gas will be diminished, and the fall of temperature will be proportional to the work done by the gas.

If in the case considered in the preceding article u be very small compared with v, the pressure upon the piston during any very small interval of time τ may be considered uniform, and the same as if the piston were at rest. The pressure upon the piston will therefore be $\frac{1}{3}\rho v^2 A$, and the work done upon it in time τ will be $\frac{1}{3}\rho v^2 A u \tau$; this will therefore be the measure of the kinetic energy lost by the gas. But the whole kinetic energy of the gas is $\frac{1}{2} M v^2$ or $\frac{1}{2} \rho v^2 A x$; and, since the absolute temperature of a gas is proportional to the kinetic energy of a given mass, it follows that if T were the absolute temperature when its volume was V, and T' the loss of temperature in expanding to the volume $V + V'$ where V' is very small compared with V, $T' : T :: u\tau : x :: V' : V$, or $T' = \frac{V'}{V} T$, where V' and T' are each indefinitely small.

In exactly the same way it may be shown that if a given mass of gas be compressed its temperature will be raised; and if the change of volume be very small, the increase of temperature will be approximately proportional to the decrease of volume.

Also, whatever be the change of volume of the gas, it may be shown, as in the case of expansion, that the increase of temperature of a given mass of the gas is proportional to the work done upon it by the agent compressing it, and that the increase of its kinetic energy is numerically equal to this amount of work.

289. If we make the piston in the case investigated in Art. 287 move through a given space with a velocity greater than that of *any* of the particles of the gas, and then suddenly come to rest, it is obvious that none of the particles of gas will impinge upon it during its motion, since they will be unable to overtake it, and when they do impinge upon it the piston will be at rest: and in this case the numerical measure of the velocity of each particle will be unaltered by impact, and therefore the kinetic energy of the gas will be unchanged. Hence also its temperature will be unchanged.

Of course we cannot experimentally make a piston move with a velocity greater than that of any of the particles of a gas, but if two chambers be separated by a diaphragm, one of them containing gas and the other a vacuum, and if the diaphragm be suddenly removed, the effect will be the same as if it were moved with infinite velocity to the extremity of the vacuum chamber. For the diaphragm substitute a tap or valve closing a pipe which connects the two otherwise closed chambers, and we have an experimental realization of the hypothesis; and we infer that if by such a contrivance a gas be allowed to expand into vacuum, its temperature will be unaltered by the expansion. This result was obtained experimentally by Dr. Joule.

290. We have seen (Art. 285) that the velocity of mean square of a gas at a given pressure is inversely proportional to the square root of its density. Suppose two gases at the

same temperature and pressure to be separated by a porous diaphragm. Then at first the number of molecules of each gas which will pass through the diaphragm into the other will be proportional to the velocity of mean square of the particles, and we have an explanation of Graham's law of diffusion, viz., that gases diffuse into one another at rates inversely proportional to the square roots of their densities.

Again, some of the gas in the first compartment having passed into the second and existing there as gas will return through the partition into the first. The amount of the first gas passing through from the first compartment to the second in the unit of time will depend on the density of the gas in the first compartment. The amount of the first gas passing from the second compartment into the first will depend on the density of this gas in the second compartment. When the density of the first gas in the second compartment is the same as in the first the amount passing per second from the first to the second will be equal to that passing in the other direction, and then it will appear as if there were no diffusion going on at all. The same will be true for the second gas, so that ultimately the two gases will be uniformly mixed in both compartments. Except that the collisions impede the process of diffusion "different gases act to one another as vacua." It is from the rate of diffusion of gases that we estimate the number of collisions or "encounters" which the molecules suffer per second, and hence deduce the number of molecules present in a unit of volume.

291. It may be shown that if two different sets of molecules are in communication the average kinetic energy of each will be finally the same. Hence the mass of a molecule of a gas must be inversely proportional to the mean square of the velocity at a given temperature, and we have seen that this is inversely proportional to the density of the gas. Hence the mass of each molecule is proportional to the density of the gas, and the number of molecules per unit volume is the same for all gases at the same temperature and pressure. This is in accordance with the fact that the chemical combining weight

302   CONCLUSION.

of a gas is proportional to its density, and with Dalton's Atomic Theory.

292. The mode in which we have treated this subject in the preceding articles is not that which we should have adopted had our object been to develop its relations to the dynamical theory of heat. We have indeed only introduced the notion of temperature because it enables us to define in a few words the condition of the gas we are considering. The definition given of temperature in treatises on heat is of course different from that which we have given, though the connection between the two is intimate. The subject of gases has been introduced here simply because it affords an example of some of the methods adopted in treating problems on elementary dynamics, the mode of investigation being merely an extension of that followed in the chain problems of Chapter IV. The student who desires a further acquaintance with the subject is referred to the articles upon it in the last chapter of Professor Clerk Maxwell's *Theory of Heat*.

293. The following method of determining the average value of $\cos^2 a$ is due to Dr. J. A. Fleming, of St. John's College.

Referring to the figure of Art. 281 and supposing it to revolve about the line AB, let σ denote the area of the zone PQ and a the angle AOP, PQ being indefinitely small. Then, denoting the radius of the sphere by v, the area of the annulus MN is $\sigma \cos a$, while $PN = v \cos a$. Hence the volume of the cylindrical shell generated by PM is $\sigma v \cos^2 a$. But the sum of all such shells makes up the volume of the hemisphere, or $\frac{2}{3} \pi v^3$. Hence $\Sigma (\sigma v \cos^2 a)$ for all values of a between 0 and $\frac{\pi}{2}$ is equal to $\frac{2}{3} \pi v^3$, and $\Sigma (\sigma \cos^2 a)$ is equal to $\frac{2}{3} \pi v^2$ since v is constant. But the sum of all such zones as σ makes up the surface of the hemisphere, or $2\pi v^2$. Hence, $\Sigma (\sigma) = 2\pi v^2$ and

$$\frac{\Sigma(\sigma \cos^2 a)}{\Sigma(\sigma)} = \frac{\frac{2}{3}\pi v^2}{2\pi v^2} = \frac{1}{3},$$

or the average value of $\cos^2 a$ for uniform distribution in space is $\frac{1}{3}$.

MISCELLANEOUS EXAMPLES.

1. The mass of each of two hammers is 30 tons, and it is moved through 3 ft. 6 in. by a constant force equal to half the weight of the hammer, the two moving in opposite directions towards one another. An inelastic mass of iron placed between the two is thus compressed so that its thickness is diminished by one inch. Supposing the pressure exerted upon the iron by each hammer to be constant throughout the compression, find its measure in pounds' weight, and the time during which the pressure acts.

Show that the diminution of thickness of the iron produced by the blow of the two hammers is twice that which either hammer would produce if the mass of iron were placed against a fixed anvil.

2. Two particles A and B, of masses $8m$ and m respectively, lie together at a point on a smooth horizontal plane, connected by a string of insensible mass which lies loose on the plane: B is projected at an elevation of 30° with a velocity equal to g; if the string become tight the instant before B reaches the plane again, and break when it has produced half the impulse it would have produced if it had not broken, and if the particle rebound at an elevation of 30°, show that the coefficient of elasticity between it and the plane is $\frac{5}{9}$.

3. A number of heavy particles are projected from the same point, (1) with the same vertical velocity, (2) with the same horizontal velocity. Show that in each case the locus of the foci of their paths is a parabola with its focus

at the point of projection and axis vertical, but in (1) the vertex is upwards, and in (2) downwards.

4. Prove that the angular velocity of a projectile about the focus of its path varies inversely as its distance from the focus.

5. Three equal particles are projected from the angular points of a triangle along the sides taken in order with velocities proportional to the sides along which they move. Prove that their centre of gravity remains at rest.

Hence show that if P, Q, R be points in the sides BC, CA, and AB respectively of the triangle ABC, such that $\dfrac{BP}{CP} = \dfrac{CQ}{AQ} = \dfrac{AR}{BR}$, then the centre of gravity of the triangle PQR coincides with that of ABC.

6. A parabola is placed with its axis vertical and vertex upwards. Prove that the square of the time of quickest descent from a given point in the axis along a chord to the curve varies as the sum of the latus rectum and the horizontal chord through that point.

7. A solid smooth cylinder of radius r lies on a smooth horizontal plane, to which it is fastened, and an inelastic sphere of radius $2r$ moves along the plane in a direction at right angles to the axis of the cylinder. Find the condition that it may pass over the cylinder.

If the sphere be elastic and the modulus of elasticity be greater than $\dfrac{1}{8}$, prove that it cannot in any case pass over the cylinder, and if e be less than $\dfrac{1}{8}$, find the condition that the sphere may, after its first ascent, fall on the top of the cylinder.

8. A particle is oscillating on the arc of a smooth cycloid whose axis is vertical. Show that the sum of its kinetic energies at any two points where the directions of motion are at right angles to each other is constant.

9. A parabolic tube is placed with its axis vertical and vertex downwards, and a particle of elasticity e starting

from the vertex with a given velocity $\sqrt{(2gh)}$ emerges, and rebounds on a horizontal plane at a depth c below the vertex. Give a geometrical construction for determining the position of the end of the tube where the particle leaves it, when the range between the first two rebounds on the horizontal plane is the greatest possible, and show that this range is $2e(h+c)$.

10. In the case of a single movable pulley, with the several portions of the string vertical, if the free end of the string pass over a fixed pulley, and if a weight be attached to the free end equal to three-fourths of the weight attached to the movable pulley, prove that the acceleration of the latter is $\frac{g}{8}$, and find the tension of the string.

If at the end of one second after the commencement of the motion the ascending weight be suddenly increased by one-half of itself, prove that its velocity will be instantaneously changed into $\frac{g}{9}$, and find the impulse of the tension of the string round the movable pulley.

11. In a system of pulleys in which all the strings are parallel, a weight of 128 lbs. is supported by a weight of 4 lbs. Suppose half a pound added to the 4 lb. weight, and the system allowed to move, find the acceleration of the 128 lb. weight, neglecting the masses of the pulleys, strings, &c.

What will be the measure of the momentum of the whole system at the end of 10 seconds, if a minute be the unit of time, 40 inches the unit of length, and $2\frac{1}{4}$ lbs. the unit of mass?

12. In a single movable pulley, when there is equilibrium, the weights hang by vertical strings:—one weight being doubled and the other halved, motion ensues. Prove that if the friction and inertia of the pulley be neglected the tension of the string will remain unaltered.

13. If a ball A move in the same direction as B, with

five times its velocity, find the ratio of the masses of the balls that A may be reduced to rest by the impact, the modulus of elasticity being $\frac{2}{3}$.

14. To one end of a string which passes over a fixed pulley is fastened a weight nW: to points $A_1, A_2,...A_n$ in the string, distant a apart, are fastened n weights each equal to W: the latter are placed close together on a horizontal plane, and motion is allowed to take place: find the velocity of the system when the last of the weights W is raised from the plane, and the time required to start the whole.

15. A smooth parabolic tube whose latus rectum is $4a$ revolves about its axis, which is vertical, with uniform angular velocity ω. Show that a heavy particle will rest in the tube in any position if $\omega = \sqrt{\dfrac{g}{2a}}$.

16. Two particles are projected from the same point at the same time with different velocities and in different directions; find the curve described by their centre of gravity.

17. Two given weights, whose masses are M and m respectively, are connected by an inextensible string which passes over a smooth fixed pulley. The system being initially at rest, determine the weight which let fall at the beginning of the motion from a point vertically above the ascending weight so as to impinge upon it will instantaneously reduce the system to rest.

18. A ball is projected from the middle point of one side of a billiard table so as to strike in succession one of the sides adjacent to it, the side opposite to it, and a ball placed in the centre of the table. Show that if a and b be the lengths of the sides of the table, and e the coefficient of elasticity of the ball and cushion, the inclination of the direction of projection to the side a of the table from which it is projected must be

$$\tan^{-1}\left\{\frac{b}{a}\cdot\frac{1+2e}{1+e}\right\}.$$

19. A string charged with $n+m+1$ equal weights fixed at equal intervals along it, and which would just rest on a smooth inclined plane with m of the weights hanging over the top, is placed on the plane with the $(m+1)^{\text{th}}$ weight just over the top; show that if a be the distance between each two adjacent weights, the velocity which the string will have acquired, at the instant the last weight slips off the plane, will be $\sqrt{nag}$.

20. Show that if a particle start from rest at one extremity of the base of a cycloidal tube, whose axis is vertical and vertex downwards, the velocity at any point is proportional to the radius of curvature at that point.

21. The tangents of the angles of a triangle ABC are in geometrical progression, $\tan B$ being the mean proportional: a ball is projected in a direction parallel to the side BC so as to strike in succession the sides AB, BC. Show that, if its course after the first impact be parallel to AC, its course after the second will be parallel to AB; and that if e be the coefficient of elasticity,

$$e^{\frac{1}{2}} + e^{-\frac{1}{2}} = \sec B.$$

22. The barrel of a rifle sighted to hit the centre of the bull's-eye, which is at the same height as the muzzle, and distant a yards from it, would be inclined at an elevation a to the horizon. Prove that if the rifle be wrongly sighted so that the elevation is $a + \theta$, θ being small compared with a, the target will be hit at a height $\dfrac{a \cos 2a}{\cos^2 a} . \theta$ above the centre of the bull's-eye.

If the range be 960 yards, the time of flight 2 seconds, and the error of elevation $1''$, the height above the centre of the bull's-eye at which the target will be hit will be nearly $\frac{1}{4}$th of an inch.

23. If the weight attached to the free end of the string in the system of pulleys in which the same string passes round each of the pulleys, be m times that which is necessary to maintain equilibrium, show that the acceleration of the ascending weight is $\dfrac{m-1}{mn+1} g$, where n

is the number of strings at the lower block, and the grooves of the pulleys are supposed perfectly smooth.

24. Two spherical balls, of elasticity e, moving in parallel directions with equal momenta, impinge; prove that if their directions of motion be opposite they will move after impact in parallel directions with equal momenta; and that these directions will be perpendicular to the original directions if their common normal at the point of impact is inclined at an angle $\sec^{-1}(1+e)^{\frac{1}{2}}$ to that direction.

25. A hollow spherical shell has a small hole at its lowest point, and any number of particles start down chords from the interior surface at the same instant, pass through the hole, and then move freely. Show that before and after passing through the hole they lie on the surface of a sphere; and determine its radius and position at any instant.

26. A string, passing over a pulley at the top of an inclined plane, connects two equal particles, one of which is placed on the plane and the other hangs freely; below the descending particle is a perfectly elastic horizontal plane; prove that if the string become stretched when this particle has reached its greatest height after the nth rebound, the inclination of the plane is $\sin^{-1}\dfrac{4n-1}{(2n-1)^2}$.

27. A sphere is fixed upon a horizontal plane; find from what point in the plane a particle must be projected, with a velocity due to falling down a vertical space equal to the diameter of the sphere, so that the focus of its path may be in the centre: show that after reflexion at the sphere, it will strike the horizontal plane at a distance from the point of projection equal to the diameter of the sphere, if the elasticity be perfect.

28. Two projectiles start from the same point at the same instant with equal velocities in the same vertical plane; prove that they will both be moving in a common tangent to their paths at the same instant after an interval of time which is the mean between the times in

the two paths from the point of projection to the point where the paths meet again.

29. If n equal masses are placed in contact in a line on a smooth table, each being connected with the next by an inelastic string of length a, and another equal mass is attached to the foremost of the n masses by a string which passes over a pulley at the edge of the table, supposing none of the n masses to leave the table before the last is set in motion, show that of the kinetic energy generated before the last is set in motion the fraction $\dfrac{n}{2(n+1)}$ is lost.

30. If for one of the weights in Attwood's machine a pulley is substituted, round which passes a string connecting two masses P, Q, which hang freely, show that if the ratio of P to Q lie between 3 and $\dfrac{1}{3}$, certain values of the other weight may be found which will keep either P or Q stationary, and that these values are to one another respectively as

$$3P-Q \text{ to } 3Q-P.$$

31. $ABCD$ is a quadrilateral inscribed in a circle: if an imperfectly elastic particle reflected at the circumference can describe the sides in order, the quadrilateral must have two of its angles right angles.

32. Particles slide from rest from a given point down straight lines to a given vertical straight line; prove that the locus of the vertices of the parabolas subsequently described is an ellipse.

33. If T be the time of descent to the vertex of a cycloid whose axis is vertical, and two particles start from one extremity of the base at an interval τ, their vertical distance when either of them is at the vertex is equal to their vertical distance when the second started, and their vertical distance is greatest after a time $\dfrac{T-\tau}{2}$.

34. An elastic ball lying at a point A on a smooth

310   EXAMPLES.

horizontal plane is driven perpendicularly against a vertical wall by the direct impact of a similar and equal ball; after rebounding from the wall at a point C, it is brought to rest by a second impact at B: show that $BC = eAC$, where e is the coefficient of elasticity between the balls.

Prove that if the stationary ball had not been in the way, the other ball would have rebounded to A in the time that was occupied in bringing the former to rest.

35. Find the latus rectum of a parabola so that when placed in a vertical plane with its axis horizontal, the least time of a particle falling from rest down a normal to the curve may be one second.

36. An inelastic particle falls from rest to a fixed inclined plane, and slides down the plane to a fixed point in it: show that the locus of the starting point is a straight line when the time to the fixed point is constant.

37. A series of smooth vertical circles touch at their highest point and particles slide down the arcs, starting from rest at the highest point: prove that the foci of the free paths of the particles lie on a straight line whose inclination to the vertical is $\tan^{-1} \dfrac{5\sqrt{5}}{8}$.

38. According to the theory of gravitation, the attraction of a sphere of radius r and density ρ, on a particle of mass m at its surface, is $\dfrac{4}{3}\pi\rho r m$. Show that if the mean density of the earth be 5·5 times that of water, the density of water in terms of the standard with reference to which ρ is measured is ·0000000067.

If 1 foot be the unit of length, find the unit of mass.

39. A string having weights w', w'' attached to its extremities, passes over two smooth fixed pulleys very close together. If a third pulley carrying a weight w be suspended by the string between the two fixed pulleys, so that all the portions of the string are vertical, and the

EXAMPLES. 311

system be left to itself, the weight w will descend with acceleration

$$g \cdot \frac{w(w'+w'')-4w'w''}{w(w'+w'')+4w'w''}.$$

40. Free particles, projected simultaneously from points on the circumference of a vertical circle towards the highest point of the circle with the velocities which would be acquired by sliding down to those points from the highest point, all reach the circumference again in the same time; and in double that time they are all in another circle of three times the radius of the former.

41. Prove that in order to produce the greatest deviation in the direction of a smooth billiard-ball of diameter a, by impact on another equal ball at rest, the former must be projected in a direction making an angle $\sin^{-1}\frac{a}{c}\sqrt{\frac{1-e}{3-e}}$ with the line (of length c) joining the two centres; e being the coefficient of elasticity.

42. If a heavy particle be projected with given velocity in vacuo, and a point be supposed to begin moving with an acceleration equal and opposite to that of gravity from the point of projection at the instant of projection, prove that at any subsequent time the particle will be moving directly away from the point, and with a velocity which in the time elapsed would have carried it over the distance between them.

43. Prove that, if the velocity of a particle be resolved into several components in one plane, its angular velocity about any fixed point in the plane is the sum of the angular velocities due to the several components.

44. If a particle move from rest with an acceleration in the direction of motion which is at first zero and increases uniformly with the time, the space described from rest in t units of time is $\frac{1}{6}ft^3$; f being the measure of the acceleration at the end of the first unit of time.

45. A string is loaded with n equal particles $A_1, A_2, \ldots A_n$ and stretched on a smooth horizontal table in a direction perpendicular to the edge; the first particle A_1 being drawn just over the edge, the system is left to the action of gravity. If the distances $c_1, c_2, \ldots$ between the particles be such that successive particles pass over the edge at equal intervals of time, and if v_r be the velocity of the system when the rth particle, A_r, is passing over the edge, and c_r be the distance between A_r and A_{r+1}, show that

$$c_r = r^2 c_1, \text{ and } v_r = r(r-1)\sqrt{\frac{gc_1}{2n}}.$$

46. The axis of a rough helix is vertical, and a small ring slides down the helix with uniform velocity. If the coefficient of friction be μ find this velocity.

47. A perfectly elastic ball is projected against one side of a smooth plane polygon, and is reflected at the other sides in succession, the polygon being such that the angle of incidence on each side is the same; find the impulse on the particle at each impact, and deduce the expression, $\dfrac{mv_2}{\rho}$, for the normal pressure on a particle moving freely on a curve under the action of no other impressed forces.

48. If a heavy particle move on a smooth curve of such form that the resultant force on the particle is equal to its weight, the radius of curvature at any point will be twice the intercept of the normal cut off by the horizontal line through the point where the velocity will be zero.

49. A thin chain is placed on a smooth horizontal table in the form of the curve in which it would hang under the action of gravity if its ends were supported, and two impulsive tensions are applied at its extremities which are to each other as the tensions at the same points in the hanging chain. Prove that the whole will move without change of form parallel to the straight line which was vertical in the hanging chain.

ANSWERS.

CHAPTER I.

EXAMINATION.

In these results π is taken to be $3\cdot1416$.

2. 1.   3. $\frac{22}{75}$.   5. $23\cdot17$...knots per hour.

6. $16\cdot49$... knots per hour.   8. $1687\frac{1}{2}$.

11. $\frac{7}{120}$ of a pound weight.   12. 3 min. $3\frac{1}{3}$ sec.

14. $\cdot5025$...   15. $151\cdot2$ lbs.' weight.

16. $\cdot306$...second.   17. $107,712,000$; $\cdot85$ H.P.

18. 396 ft.; 6 seconds.   19. $\frac{1}{3}$ lb. $\frac{2}{3}$ ft. $\frac{1}{12}$ sec.

20. $6\cdot514$...ins.   21. $3\cdot0237$...ins. $\cdot0112$...secs.

22. 50 lbs. $6\cdot4$ ft. per second.   23. $14\frac{41}{56}$ ft. $\cdot339$...secs.

24. $2036\frac{4}{11}$ H.P.

EXAMPLES.

1. $17\cdot338$...; $1525\cdot7$...   2. 66; 60.

3. $1582\cdot62$.   4. 20 knots per hour.

5. 40 miles per hour.   6. 7 : 8.

7. 480 ft. per second. 96,000 units of momentum.

8. $\frac{1}{14} g$.   9. $\frac{12}{175}$ foot second units.

314 ANSWERS.

10. 1 hr. 23 min. 20 sec.
11. 25,980·7...foot-pounds.   ·787... h.p.
12. 1,543,238·68...foot-tons.   15. 384 h.p.
16. $25\frac{35}{27}$ lbs.' weight per square inch.   17. $5500\frac{143}{160}$ h.p.
19. 44 ft. per second.   22. $g \tan a$.
24. 9·68 feet.   25. 1,228,800 lbs.
26. 192,000.   27. ·605 ft.   ·1375 sec.
28. $\frac{a^2r^2}{bt^2}$ ft. $\frac{ar^2}{bt}$ secs.   29. 60 : 1.   12 : 1.
30. 11 seconds.   31. $53\frac{1}{3}$ lbs.
32. 40 ft. per second.   20 ; 72,000.   33. $\frac{a^2}{32t^2}$ ft., $\frac{a}{32t}$ seconds.
37. 90,000,000.   38. $1592\frac{8}{9}$.
39. 6,055,723,745,·28 tons.   12947 ft. nearly.   1190·2...days.
41. $\frac{1}{100}$ ft., $\frac{1}{800}$ oz., $\frac{1}{100}$ sec.   42. $\frac{99}{1300}$.
44. 600 feet.   1200 lbs. $7\frac{1}{2}$ seconds.

CHAPTER II.

EXAMINATION.

1. 88 ft.   2. 256 feet. $4\frac{8}{35}$ secs.
3. 2560.   4. $8\sqrt{\frac{2}{33}}$ ft. per second.
5. 5 seconds.   6. $8\sqrt{5}$ ft. per second. $\frac{1}{2}\sqrt{5}$ seconds.
7. 1·424...seconds.   28·08 ft. per second, nearly.
9. ·9463...sec.   10. $\frac{1}{2}\sqrt{\frac{3}{\cos^2 15°}}$ sec.
11. $160m$ foot-second units.   12. 8.

ANSWERS.

EXAMPLES.

2. $\dfrac{ft}{v}$ seconds. $2\sigma \dfrac{ft^2}{v^2 r^2}$ feet.   4. 5192·3... feet.

5. $28\tfrac{1}{2}\tfrac{4}{7}$ miles.   6. $4656\tfrac{2}{3}$ feet.

7. $\dfrac{2167}{5040}$. $10038\tfrac{11}{47}$ ft.   706·4... ft.

8. $19\tfrac{3}{13}$ ft. per second.   22·326... ft. along the plane.

9. 15 : 1456.   10. 17·05... pounds' weight.

12. 2 secs. or 8 secs.; $3g$.

13. They meet $\dfrac{v}{g} \cdot \dfrac{\sqrt{2}+1}{2\sqrt{2}}$ seconds after projecting the second ball, and at a height above the point of projection equal to $\dfrac{v^2}{g} \cdot \dfrac{5+2\sqrt{2}}{16}$.

14. 2 seconds. 3 ft.   15. $4\tfrac{1}{2}$ seconds.

16. Equal to the weight of 19 cwt. $10\tfrac{3}{9}$ lbs.   19. $32 : g$; $21\tfrac{9}{11}$.

20. The *resultant* force on the particle must in the first case be 4 times its weight, and in the second case 2560 times its weight.

21. $\dfrac{ft}{b}$ seconds. $\dfrac{2af}{b^2}$ feet.   22. $\tfrac{1}{4}gt$; $\tfrac{3}{4}W$. $W\dfrac{\sqrt{3}}{8}$.

23. 2:1.   24. 3600 m.

27. $2\tfrac{4}{7}P$. $\sqrt{\dfrac{28l}{5g}}$ seconds, where l is the length of the plane.

28. n^2f units of length per second. $nf\left\{\tfrac{2}{3}n^2 + \tfrac{1}{2}n + \tfrac{1}{3}\right\}$ feet.

42. 100 yards.   43. 14175 feet, 945 seconds, 88 tons.

44. 800 feet. 10 seconds.   46. $11\tfrac{3}{14}$ tons.

48. 28 : 165.   49. 9·33 H.P.; 4·4 H.P.

CHAPTER III.

EXAMINATION.

1. 25 feet.

3. $900\sqrt{99}$ feet.

4. 625 feet.   312·5 feet.

6. The elevation of projection is $\tan^{-1}\dfrac{20\pm\sqrt{311}}{3}$.

7. Nearly 66 feet from the point of projection.

8. The direction makes an angle of 45° with the lines of slope.

9. 121 feet.

10. ·218... $\cos\lambda$ inches to the East of the foot of the mast, where λ is the latitude of the place.

EXAMPLES.

1. 618·01 feet per second; $\tan^{-1}\dfrac{512}{1125}$.   2. 1 : 2.

3. Inclined 45° to the horizon. 158·6... feet or 372·3... feet.

5. Inclined 45° to the horizon. Inclined at an angle $\tan^{-1}\dfrac{1}{\sqrt{2}}$ to the horizon.

6. Between 15° and 15° 42′ 43″ nearly, or between 75° and 74° 56′ 36″ nearly.

13. The $(N+1)^{\text{th}}$ stair from the top where N is the greatest integer in $\dfrac{4na}{c}$.

16. 30°.   23. $8\sqrt{10}$ feet per second.

28. Velocity $= \sqrt{g\{2c+(a\sim b)\}}$. Elevation $= \tan^{-1}\dfrac{a+b}{a\sim b}$, and we must have $c = \dfrac{ab}{a\sim b}$.

35. Time after projection $= \dfrac{u}{g\sin a}$; velocity $= u \cot a$; where a is the elevation and u the velocity of projection.

37. Nearly 26000 square miles.

ANSWERS. 317

CHAPTER IV.

EXAMINATION.

2. $40\sqrt{2}$.

3. 1·657... units of force. 9·6834... feet per second.

5. The direction of motion makes an angle $\tan^{-1}\dfrac{2}{2\sqrt{3}-1}$ with the plane, and the velocity is 6·478... feet per second.

6. The directions of motion after impact make with the line joining the centres at the instant of impact the angles $\tan^{-1}\dfrac{32}{15\sqrt{3}}$ and $\tan^{-1}\dfrac{16}{\sqrt{3}}$ respectively. The velocities are $1\cdot 288...v$ and $\cdot 5029...v$ respectively.

7. ·8398...mv absolute units.

8. 7·5 feet. 54·772... feet.

10. 22·82... pounds' weight.

11. $240\,(1-e^2)$ units of energy.

EXAMPLES.

1. $15\frac{1}{3}$, the balls moving at first in the same direction.

2. 484 absolute units.

3. $\dfrac{1200}{19201}$ feet per second.

4. $14\frac{2}{7}$ feet per second; $11\frac{9}{56}$ feet.

5. $26\frac{2}{7}$ and $51\frac{3}{7}$ ft. per second; $2142\frac{6}{7}$ absolute units of impulse.

8. The mass of the moving ball must be e times that of the ball at rest.

9. The velocity is equal to $\dfrac{3}{4}\sqrt{5ga}$, where a represents the side of the square.

ANSWERS.

10. $\frac{2}{25}$.

11. 8 feet per second; $\frac{1}{4}$ second.

14. $\dfrac{mm'(u-v)}{m+m'} \cdot \dfrac{3}{2\sqrt{2}}$.

15. $\dfrac{mu}{2M+m}(1+e)$.

17. $2:1$. v feet, where v represents the velocity of each before impact.

18. If θ denote the elevation of projection and a the length of a side of the square, $v = \sqrt{\dfrac{ga}{2\cos\theta(\sin\theta - \cos\theta)}}$.

21. $\pi - 2\tan^{-1}e \sqrt{\dfrac{e}{1+e+e^2}}$.

28. The direction makes an angle $\tan^{-1}\dfrac{3-2\sqrt{2}}{3\sqrt{3}+2\sqrt{2}}$ with the common tangent, and the velocity is ·80... times the velocity of either ball before impact.

30. The first ball will fall at a distance $g\sqrt{3}$ from its point of projection, and the second at the point of projection of the first.

37. $1 - e - e^2 + e^{n+1} : 1 - e^2$.

38. At one extremity of the latus rectum.

42. $v = \sqrt{\dfrac{ag}{2}}$, where a is the radius of the sphere.

45. It makes an angle $\dfrac{\pi}{2} - a$ with the plane through the point of projection and the intersection of the walls.

48. If m be the mass of the chain, and M that of each bucket, the pressure is $\dfrac{M+3m}{M+2m}mg$

52. The range is equal to $8QN$. The locus is a straight line through the fixed point inclined to the plane at an angle whose tangent is one-ninth of the cotangent of the inclination of the plane.

ANSWERS. 319

54. They will continue to descend with an acceleration $\dfrac{2-\sqrt{3}}{4}g$.

56. Inversely as their masses.

65. $e = \dfrac{M}{M+2m}$, where M is the mass of the moving sphere, and m that of the sphere initially at rest.

66. $e = \dfrac{4-n^2}{n^2}$; n must lie between $\sqrt{2}$ and 2.

CHAPTER V.

Examination.

1. $v = \sqrt{2gs}$, if s be the focal distance of the point.

2. $4\tfrac{10}{11}$ nearly.

4. It leaves the surface after sliding through a vertical distance equal to one-third the radius of the cylinder.

5. The inclination of the string to the vertical is $\cos^{-1}\dfrac{g}{108\pi^2}$, and the tension is equal to $2160\pi^2$ absolute units of force.

6. $4\sqrt{10}$ feet per second.   8. Zero.   9. 19·77...inches.

10. Change of tension $= \dfrac{w}{3\sqrt{6}}$.   11. $ma^2\omega^2$.

12. $\dfrac{2}{7}g$. $1\tfrac{3}{7}$ lbs' weight.

Examples.

1. 134·1...   3. $\cos^{-1}\dfrac{54}{125\pi^2}$.   4. $\tan^{-1}\dfrac{1089}{4072}$.

5. 179·18... pounds' weight.

6. 1 : 1·00111...   8. 16·12...feet.

9. 1206·2925 yards.   11. $40\tfrac{40}{59}$ inches.

ANSWERS.

14. $12\frac{3}{11}$ nearly.   15. 1·1 feet nearly.   17. $\sqrt{gr}$.

18. Distance from vertex $= \dfrac{g}{\omega^2} \cdot \dfrac{\cos a}{\sin^2 a}$.   20. ·018...

23. $\dfrac{18\pi a}{v}$; $\dfrac{mv^2}{6a}$; $\dfrac{mv^2}{a}$.

24. Inclination to the vertical $= \tan^{-1}\dfrac{\pi\sqrt{1+4\pi^2}}{32}$. Magnitude $=$ 198·08 lbs.' weight nearly.

30. Acceleration of wedge $= g \cdot \dfrac{4\sqrt{2}+2\cos 15°}{9}$, or, $g \cdot \dfrac{1+3\sqrt{3}}{3\sqrt{6}}$.

Pressure between weight and wedge $= \dfrac{40\cos 15° - 10\sqrt{2}}{9}$ lbs.' weight, or, $\dfrac{10\sqrt{2}}{3\sqrt{3}} g$ poundals.

Pressure between wedge and plane $= \dfrac{45\sqrt{2}+20\sqrt{3}\cos 15° - 5\sqrt{6}}{9}$ lbs.' weight, or, $\dfrac{20\sqrt{2}}{3} g$ poundals.

Acceleration of weight $= g \cdot \dfrac{\sqrt{17-6\sqrt{3}}}{3\sqrt{3}}$.

31. Acceleration of weight $= \dfrac{2}{35}$.

Space $= 102\frac{2}{3}$ feet.

THE END.

November 1889.

A CLASSIFIED LIST

OF

EDUCATIONAL WORKS

PUBLISHED BY

GEORGE BELL & SONS.

Cambridge Calendar. Published Annually (*August*). 6s. 6d.
Student's Guide to the University of Cambridge. 6s. 6d.
Oxford: Its Life and Schools. 7s. 6d.
The Schoolmaster's Calendar. Published Annually (*December*). 1s.

BIBLIOTHECA CLASSICA.

A Series of Greek and Latin Authors, with English Notes, edited by eminent Scholars. 8vo.

⁎⁎ *The Works with an asterisk (*) prefixed can only be had in the Sets of 26 Vols.*

Aeschylus. By F. A. Paley, M.A., LL.D. 8s.
Cicero's Orations. By G. Long, M.A. 4 vols. 32s.
Demosthenes. By R. Whiston, M.A. 2 vols. 10s.
Euripides. By F. A. Paley, M.A., LL.D. 3 vols. 24s.
Homer. By F. A. Paley, M.A., LL.D. The Iliad, 2 vols. 14s.
Herodotus. By Rev. J. W. Blakesley, B.D. 2 vols. 12s.
Hesiod. By F. A. Paley, M.A., LL.D. 5s.
Horace. By Rev. A. J. Macleane, M.A. 8s.
Juvenal and Persius. By Rev. A. J. Macleane, M.A. 6s.
Plato. By W. H. Thompson, D.D. 2 vols. 5s. each.
Sophocles. Vol. I. By Rev. F. H. Blaydes, M.A. 8s.
—— Vol. II. F. A. Paley, M.A., LL.D. 6s.
*****Tacitus: The Annals.** By the Rev. P. Frost. 8s.
*****Terence.** By E. St. J. Parry, M.A. 8s.
Virgil. By J. Conington, M.A. Revised by Professor H. Nettleship.
3 vols. 10s. 6d. each.
An Atlas of Classical Geography; 24 Maps with coloured Outlines. Imp. 8vo. 6s.

GRAMMAR-SCHOOL CLASSICS.

A Series of Greek and Latin Authors, with English Notes.
Fcap. 8vo.

Cæsar: De Bello Gallico. By George Long, M.A. 4s.
―――― Books I.-III. For Junior Classes. By G. Long, M.A. 1s. 6d.
―――― Books IV. and V. 1s. 6d. Books VI. and VII. 1s. 6d.
Catullus, Tibullus, and Propertius. Selected Poems. With Life. By Rev. A. H. Wratislaw. 2s. 6d.
Cicero: De Senectute, De Amicitia, and Select Epistles. By George Long, M.A. 3s.
Cornelius Nepos. By Rev. J. F. Macmichael. 2s.
Homer: Iliad. Books I.-XII. By F. A. Paley, M.A., LL.D. 4s. 6d. Also in 2 parts, 2s. 6d. each.
Horace: With Life. By A. J. Macleane, M.A. 3s. 6d. In 2 parts, 2s. each.
Juvenal: Sixteen Satires. By H. Prior, M.A. 3s. 6d.
Martial: Select Epigrams. With Life. By F. A. Paley, M.A., LL.D. 4s. 6d.
Ovid: the Fasti. By F. A. Paley, M.A., LL.D. 3s. 6d. Books I. and II. 1s. 6d. Books III. and IV. 1s. 6d.
Sallust: Catilina and Jugurtha. With Life. By G. Long, M.A. and J. G. Frazer. 3s. 6d., or separately, 2s. each.
Tacitus: Germania and Agricola. By Rev. P. Frost. 2s. 6d.
Virgil: Bucolics, Georgics, and Æneid, Books I.-IV. Abridged from Professor Conington's Edition. 4s. 6d.—Æneid, Books V.-XII. 4s. 6d. Also in 9 separate Volumes, as follows, 1s. 6d. each:—Bucolics—Georgics, I. and II.—Georgics, III. and IV.—Æneid, I. and II.—Æneid, III. and IV.—Æneid, V. and VI.—Æneid, VII. and VIII.—Æneid, IX. and X.—Æneid, XI. and XII.
Xenophon: The Anabasis. With Life. By Rev. J. F. Macmichael. 3s. 6d. Also in 4 separate volumes, 1s. 6d. each:—Book I. (with Life, Introduction, Itinerary, and Three Maps)—Books II. and III.—IV. and V.—VI. and VII.
―――― The Cyropædia. By G. M. Gorham, M.A. 3s. 6d. Books I. and II. 1s. 6d.—Books V. and VI. 1s. 6d.
―――― Memorabilia. By Percival Frost, M.A. 3s.
A Grammar-School Atlas of Classical Geography, containing Ten selected Maps. Imperial 8vo. 3s.

Uniform with the Series.

The New Testament, in Greek. With English Notes, &c. By Rev. J. F. Macmichael. 4s. 6d. In 5 parts, The Four Gospels and the Acts. Sewed, 6d. each.

Educational Works. 3

CAMBRIDGE GREEK AND LATIN TEXTS.

Aeschylus. By F. A. Paley, M.A., LL.D. 2s.
Cæsar: De Bello Gallico. By G. Long, M.A. 1s. 6d.
Cicero: De Senectute et De Amicitia, et Epistolæ Selectæ. By G. Long, M.A. 1s. 6d.
Ciceronis Orationes. In Verrem. By G. Long, M.A. 2s. 6d.
Euripides. By F. A. Paley, M.A., LL.D. 3 vols. 2s. each.
Herodotus. By J. G. Blakesley, B.D. 2 vols. 5s.
Homeri Ilias. I.-XII. By F. A. Paley, M.A., LL.D. 1s. 6d.
Horatius. By A. J. Macleane, M.A. 1s. 6d.
Juvenal et Persius. By A. J. Macleane, M.A. 1s. 6d.
Lucretius. By H. A. J. Munro, M.A. 2s.
Sallusti Crispi Catilina et Jugurtha. By G. Long, M.A. 1s. 6d.
Sophocles. By F. A. Paley, M.A., LL.D. 2s. 6d.
Terenti Comœdiæ. By W. Wagner, Ph.D. 2s.
Thucydides. By J. G. Donaldson, D.D. 2 vols. 4s.
Virgilius. By J. Conington, M.A. 2s.
Xenophontis Expeditio Cyri. By J. F. Macmichael, B.A. 1s. 6d.
Novum Testamentum Græce. By F. H. Scrivener, M.A., D.C.L. 4s. 6d. An edition with wide margin for notes, half bound, 12s. ELITIO MAJOR, with additional Readings and References. 7s. 6d. *See page* 14.

CAMBRIDGE TEXTS WITH NOTES.

A Selection of the most usually read of the Greek and Latin Authors, Annotated for Schools. Edited by well-known Classical Scholars. Fcap. 8vo. 1s. 6d. each, with exceptions.

'Dr. Paley's vast learning and keen appreciation of the difficulties of beginners make his school editions as valuable as they are popular. In many respects he sets a brilliant example to younger scholars.'—*Athenæum.*

'We hold in high value these handy Cambridge texts with Notes.'—*Saturday Review.*

Aeschylus. Prometheus Vinctus.—Septem contra Thebas.—Agamemnon.—Persae.—Eumenides.—Choephoroe. By F.A. Paley, M.A., LL.D.
Euripides. Alcestis.—Medea.—Hippolytus.—Hecuba.—Bacchae. —Ion. 2s.—Orestes. — Phoenissæ.—Troades.—Hercules Furens.—Andromache.—Iphigenia in Tauris.—Supplices. By F. A. Paley, M.A., LL.D.
Homer. Iliad. Book I. By F. A. Paley, M.A., LL.D. 1s.
Sophocles. Oedipus Tyrannus.—Oedipus Coloneus.—Antigone. —Electra—Ajax. By F. A. Paley, M.A., LL.D.
Xenophon. Anabasis. In 6 vols. By J. E. Melhuish, M.A., Assistant Classical Master at St. Paul's School.
——— Hellenics, Book II. By L. D. Dowdall, M.A., B.D. 2s.
——— Hellenics. Book I. By L. D. Dowdall, M.A., B.D.
[*In the press.*
Cicero. De Senectute, De Amicitia and Epistolæ Selectæ. By G. Long, M.A.
Ovid. Fasti. By F. A. Paley, M.A. LL.D. In 3 vols., 2 books in each. 2s. each vol.

Ovid. Selections. Amores, Tristia, Heroides, Metamorphoses.
By A. J. Macleane, M.A.
Terence. Andria.—Hauton Timorumenos.—Phormio.—Adelphoe.
By Professor Wagner, Ph.D.
Virgil. Professor Conington's edition, abridged in 12 vols.
Others in preparation.

PUBLIC SCHOOL SERIES.
A Series of Classical Texts, annotated by well-known Scholars. Cr. 8vo.

Aristophanes. The Peace. By F. A. Paley, M.A., LL.D. 4s. 6d.
—— The Acharnians. By F. A. Paley, M.A., LL.D. 4s. 6d.
—— The Frogs. By F. A. Paley, M.A., LL.D. 4s. 6d.
Cicero. The Letters to Atticus. Bk. I. By A. Pretor, M.A. 4s. 6d.
Demosthenes de Falsa Legatione. By R. Shilleto, M.A. 6s.
—— The Law of Leptines. By B. W. Beatson, M.A. 3s. 6d.
Livy. Book XXI. Edited, with Introduction, Notes, and Maps, by the Rev. L. D. Bowdall, M.A., B.D. 3s. 6d.
—— Book XXII. Edited, &c., by Rev. L. D. Dowdall, M.A., B.D. 3s. 6d.
Plato. The Apology of Socrates and Crito. By W. Wagner, Ph.D. 10th Edition. 3s. 6d. Cheap Edition, limp cloth, 2s. 6d.
—— The Phædo. 9th Edition. By W. Wagner, Ph.D. 5s. 6d.
—— The Protagoras. 4th Edition. By W. Wayte, M.A. 4s. 6d.
—— The Euthyphro. 3rd Edition. By G. H. Wells, M.A. 3s.
—— The Euthydemus. By G. H. Wells, M.A. 4s.
—— The Republic. Books I. & II. By G. H. Wells, M.A. 3rd Edition. 5s. 6d.
Plautus. The Aulularia. By W. Wagner, Ph.D. 3rd Edition. 4s. 6d.
—— The Trinummus. By W. Wagner, Ph.D. 3rd Edition. 4s. 6d.
—— The Menaechmei. By W. Wagner, Ph.D. 2nd Edit. 4s. 6d.
—— The Mostellaria. By Prof. E. A. Sonnenschein. 5s.
—— The Rudens. Edited by Prof. E. A. Sonnenschein.
[*In the press.*
Sophocles. The Trachiniæ. By A. Pretor, M.A. 4s. 6d.
Sophocles. The Oedipus Tyrannus. By B. H. Kennedy, D.D. 5s.
Terence. By W. Wagner, Ph.D. 2nd Edition. 7s. 6d.
Theocritus. By F. A. Paley, M.A., LL.D. 2nd Edition. 4s. 6d.
Thucydides. Book VI. By T. W. Dougan, M.A., Fellow of St. John's College, Cambridge. 3s. 6d.
Others in preparation.

CRITICAL AND ANNOTATED EDITIONS.
Aristophanis Comœdiæ. By H. A. Holden, LL.D. 8vo. 2 vols.
Notes, Illustrations, and Maps. 23s. 6d. Plays sold separately.
Cæsar's Seventh Campaign in Gaul, B.C. 52. By Rev. W. C. Compton, M.A., Assistant Master, Uppingham School. Crown 8vo. 4s.

Educational Works. 5

Calpurnius Siculus. By C. H. Keene, M.A. Crown 8vo. 6s.
Catullus. A New Text, with Critical Notes and Introduction
by Dr. J. P. Postgate. Japanese vellum. Foolscap 8vo. 3s.
Corpus Poetarum Latinorum. Edited by Walker. 1 vol. 8vo. 18s.
Horace. Quinti Horatii Flacci Opera. By H. A. J. Munro, M.A.
Large 8vo. 10s. 6d.
Livy. The first five Books. By J. Prendeville. 12mo. roan, 5s.
Or Books I.-III. 3s. 6d. IV. and V. 3s. 6d. Or the five Books in separate
vols. 1s. 6d. each.
Lucan. The Pharsalia. By C. E. Haskins, M.A., and W. E.
Heitland, M.A. Demy 8vo. 14s.
Lucretius. With Commentary by H. A. J. Munro. 4th Edition.
Vols. I. and II. Introduction, Text, and Notes. 18s. Vol. III. Translation. 6s.
Ovid. P. Ovidii Nasonis Heroides XIV. By A. Palmer, M.A. 8vo. 6s.
——— P. Ovidii Nasonis Ars Amatoria et Amores. By the Rev.
H. Williams, M.A. 3s. 6d.
——— Metamorphoses. Book XIII. By Chas. Haines Keene, M.A.
2s. 6d.
——— Epistolarum ex Ponto Liber Primus. By C.H.Keene,M.A. 3s.
Propertius. Sex Aurelii Propertii Carmina. By F. A. Paley, M.A.,
LL.D. 8vo. Cloth, 5s.
——— Sex Propertii Elegiarum. Libri IV. Recensuit A. Palmer,
Collegii Sacrosanctæ et Individuæ Trinitatis juxta Dublinum Socius.
Fcap. 8vo. 3s. 6d.
Sophocles. The Oedipus Tyrannus. By B. H. Kennedy, D.D.
Crown 8vo. 8s.
Thucydides. The History of the Peloponnesian War. By Richard
Shilleto, M.A. Book I. 8vo. 6s. 6d. Book II. 8vo. 5s. 6d.

LOWER FORM SERIES.
With Notes and Vocabularies.

Eclogæ Latinæ; or, First Latin Reading-Book, with English Notes
and a Dictionary. By the late Rev. P. Frost, M.A. New Edition. Fcap.
8vo. 1s. 6d.
Latin Vocabularies for Repetition. By A. M. M. Stedman, M.A.
2nd Edition, revised. Fcap. 8vo. 1s. 6d.
Easy Latin Passages for Unseen Translation. By A. M. M.
Stedman, M.A. Fcap. 8vo. 1s. 6d.
Virgil's Æneid. Book I. Abridged from Conington's Edition by
Rev. J. G. Sheppard, D.C.L. With Vocabulary by W. F. R. Shilleto.
1s. 6d. [*Now ready.*
Cæsar de Bello Gallico. Books I., II. and III. With Notes by
George Long, M.A., and Vocabulary by W. F. R. Shilleto. 1s. 6d. each.
Tales for Latin Prose Composition. With Notes and Vocabulary. By G. H. Wells, M.A. 2s.
A Latin Verse-Book. An Introductory Work on Hexameters and
Pentameters. By the late Rev. P. Frost, M.A. New Edition. Fcap. 8vo.
2s. Key (for Tutors only), 5s.
Analecta Græca Minora, with Introductory Sentences, English
Notes, and a Dictionary. By the late Rev. P. Frost, M.A. New Edition.
Fcap. 8vo. 2s.
Greek Testament Selections. 2nd Edition, enlarged, with Notes
and Vocabulary. By A. M. M. Stedman, M.A. Fcap. 8vo. 2s. 6d.

LATIN AND GREEK CLASS-BOOKS.
(*See also Lower Form Series.*)

Faciliora. An Elementary Latin Book on a new principle. By the Rev. J. L. Seager, M.A. 2s. 6d.

First Latin Lessons. By A. M. M. Stedman. 1s.

Easy Latin Exercises, for Use with the Revised Latin Primer and Shorter Latin Primer. By A. M. M. Stedman, M.A. (*Issued with the consent of the late Dr. Kennedy.*) Crown 8vo. 2s. 6d.

Miscellaneous Latin Exercises. By A. M. M. Stedman, M.A. Fcap. 8vo. 1s. 6d.

A Latin Primer. By Rev. A. C. Clapin, M.A. 1s.

Auxilia Latina. A Series of Progressive Latin Exercises. By M. J. B. Baddeley, M.A. Fcap. 8vo. Part I. Accidence. 3rd Edition, revised. 2s. Part II. 4th Edition, revised. 2s. Key to Part II. 2s. 6d.

Scala Latina. Elementary Latin Exercises. By Rev. J. W. Davis, M.A. New Edition, with Vocabulary. Fcap. 8vo. 2s. 6d.

Passages for Translation into Latin Prose. By Prof. H. Nettleship, M.A. 3s. Key (for Tutors only), 4s. 6d.

Latin Prose Lessons. By Prof. Church, M.A. 9th Edition. Fcap. 8vo. 2s. 6d.

Analytical Latin Exercises. By C. P. Mason, B.A. 4th Edit. Part I., 1s. 6d. Part II., 2s. 6d.

By T. COLLINS, M.A., Head Master of the Latin School, Newport, Salop.

Latin Exercises and Grammar Papers. 6th Edit. Fcap. 8vo. 2s. 6d.

Unseen Papers in Latin Prose and Verse. With Examination Questions. 4th Edition. Fcap. 8vo. 2s. 6d.

—— **in Greek Prose and Verse.** With Examination Questions. 3rd Edition. Fcap. 8vo. 3s.

Easy Translations from Nepos. Cæsar, Cicero, Livy, &c., for Retranslation into Latin. With Notes. 2s.

By A. M. M. STEDMAN, M.A., Wadham College, Oxford.

Latin Examination Papers in Grammar and Idiom. 2nd Edition. Crown 8vo. 2s. 6d. Key (for Tutors and Private Students only), 6s.

Greek Examination Papers in Grammar and Idiom. 2s. 6d.

By the REV. P. FROST, M.A., St. John's College, Cambridge.

Materials for Latin Prose Composition. By the late Rev. P. Frost, M.A. New Edition. Fcap. 8vo. 2s. Key (for Tutors only), 4s.

Materials for Greek Prose Composition. New Edit. Fcap. 8vo. 2s. 6d. Key (for Tutors only), 5s.

Florilegium Poeticum. Elegiac Extracts from Ovid and Tibullus. New Edition. With Notes. Fcap. 8vo. 2s.

By H. A. HOLDEN, LL.D., formerly Fellow of Trinity Coll., Camb.

Foliorum Silvula. Part I. Passages for Translation into Latin Elegiac and Heroic Verse. 10th Edition. Post 8vo. 7s. 6d.

—— Part II. Select Passages for Translation into Latin Lyric and Comic Iambic Verse. 3rd Edition. Post 8vo. 5s.

Folia Silvulæ, sive Eclogæ Poetarum Anglicorum in Latinum et Græcum conversæ. 8vo. Vol. II. 4s. 6d.

Foliorum Centuriæ. Select Passages for Translation into Latin and Greek Prose. 10th Edition. Post 8vo. 8s.

Educational Works. 7

Scala Græca: a Series of Elementary Greek Exercises. By Rev. J. W. Davis, M.A., and R. W. Baddeley, M.A. 3rd Edition. Fcap. 8vo. 2s. 6d.
Greek Verse Composition. By G. Preston, M.A. 5th Edition. Crown 8vo. 4s. 6d.
Greek Particles and their Combinations according to Attic Usage. A Short Treatise. By F. A. Paley, M.A., LL.D. 2s. 6d.
Rudiments of Attic Construction and Idiom. By the Rev. W. C. Compton, M.A., Assistant Master at Uppingham School. 3s.
Anthologia Græca. A Selection of Choice Greek Poetry, with Notes. By F. St. John Thackeray. 4th and Cheaper Edition. 16mo. 4s. 6d.
Anthologia Latina. A Selection of Choice Latin Poetry, from Nævius to Boëthius, with Notes. By Rev. F. St. John Thackeray. Revised and Cheaper Edition. 16mo. 4s. 6d.

TRANSLATIONS, SELECTIONS, &c.

*** Many of the following books are well adapted for School Prizes.
Aeschylus. Translated into English Prose by F. A. Paley, M.A., LL.D. 2nd Edition. 8vo. 7s. 6d.
────── Translated into English Verse by Anna Swanwick. 4th Edition. Post 8vo. 5s.
Horace. The Odes and Carmen Sæculare. In English Verse by J. Conington, M.A. 10th edition. Fcap. 8vo. 5s. 6d.
────── The Satires and Epistles. In English Verse by J. Conington, M.A. 7th edition. 6s. 6d.
────── Odes. Englished and Imitated by various hands. 1s. 6d.
Plato. Gorgias. Translated by E. M. Cope, M.A. 8vo. 2nd Ed. 7s.
────── Philebus. Trans. by F. A. Paley, M.A., LL.D. Sm. 8vo. 4s.
────── Theætetus. Trans. by F. A. Paley, M.A., LL.D. Sm. 8vo. 4s.
────── Analysis and Index of the Dialogues. By Dr. Day. Post 8vo. 5s.
Sophocles. Oedipus Tyrannus. By Dr. Kennedy. 1s.
────── The Dramas of. Rendered into English Verse by Sir George Young, Bart., M.A. 8vo. 12s. 6d.
Theocritus. In English Verse, by C. S. Calverley, M.A. New Edition, revised. Crown 8vo. 7s. 6d.
Translations into English and Latin. By C. S. Calverley, M.A. Post 8vo. 7s. 6d.
Translations into English, Latin, and Greek. By R. C. Jebb, Litt.D., H. Jackson, Litt.D., and W. E. Currey, M.A. Second Edition. 8s.
Extracts for Translation. By R. C. Jebb, Litt. D., H. Jackson, Litt.D., and W. E. Currey, M.A. 4s. 6d.
Between Whiles. Translations by Rev. B. H. Kennedy, D.D. 2nd Edition, revised. Crown 8vo. 5s.
Sabrinae Corolla in Hortulis Regiae Scholae Salopiensis Contexuerunt Tres Viri Floribus Legendis. Fourth Edition, thoroughly Revised and Rearranged. With many new Pieces and an Introduction.
[*Ready immediately.*

REFERENCE VOLUMES.

A Latin Grammar. By Albert Harkness. Post 8vo. 6s.
────── By T. H. Key, M.A. 6th Thousand. Post 8vo. 8s.
A Short Latin Grammar for Schools. By T. H. Key, M.A. F.R.S. 16th Edition. Post 8vo. 3s. 6d.

A Guide to the Choice of Classical Books. By J. B. Mayor, M.A. 3rd Edition, with a Supplementary List. Crown 8vo. 4s. 6d. Supplementary List separately, 1s. 6d.
The Theatre of the Greeks. By J. W. Donaldson, D.D. 8th Edition. Post 8vo. 5s.
Keightley's Mythology of Greece and Italy. 4th Edition. 5s.

CLASSICAL TABLES.

Latin Accidence. By the Rev. P. Frost, M.A. 1s.
Latin Versification. 1s.
Notabilia Quædam; or the Principal Tenses of most of the Irregular Greek Verbs and Elementary Greek, Latin, and French Construction. New Edition. 1s.
Richmond Rules for the Ovidian Distich, &c. By J. Tate, M.A. 1s.
The Principles of Latin Syntax. 1s.
Greek Verbs. A Catalogue of Verbs, Irregular and Defective. By J. S. Baird, T.C.D. 8th Edition. 2s. 6d.
Greek Accents (Notes on). By A. Barry, D.D. New Edition. 1s.
Homeric Dialect. Its Leading Forms and Peculiarities. By J. S. Baird, T.C.D. New Edition, by W. G. Rutherford, LL.D. 1s.
Greek Accidence. By the Rev. P. Frost, M.A. New Edition. 1s.

CAMBRIDGE MATHEMATICAL SERIES.

Arithmetic for Schools. By C. Pendlebury, M.A. 3rd Edition, revised and stereotyped, with or without answers, 4s. 6d. Or in two parts, 2s. 6d. each.
    EXAMPLES (nearly 8000), without answers, in a separate vol. 3s.
    In use at St. Paul's, Winchester, Charterhouse, Merchant Taylors', Christ's Hospital, Sherborne, Shrewsbury, and at many other Schools and Colleges.
Algebra. Choice and Chance. By W. A. Whitworth, M.A. 4th Edition. 6s.
Euclid. Books I.-VI. and part of Books XI. and XII. By H. Deighton. 4s. 6d. Key (for Tutors only), 5s. Books I. and II., 2s.
Euclid. Exercises on Euclid and in Modern Geometry. By J. McDowell, M.A. 3rd Edition. 6s.
Trigonometry. Plane. By Rev. T. Vyvyan, M.A. 3rd Edit. 3s. 6d.
Geometrical Conic Sections. By H. G. Willis, M.A. Manchester Grammar School. 5s.
Conics. The Elementary Geometry of. 6th Edition, revised and enlarged. By C. Taylor, D.D. 4s. 6d.
Solid Geometry. By W. S. Aldis, M.A. 4th Edit. revised. 6s.
Geometrical Optics. By W. S. Aldis, M.A. 3rd Edition. 4s.
Rigid Dynamics. By W. S. Aldis, M.A. 4s.
Elementary Dynamics. By W. Garnett, M.A., D.C.L. 5th Ed. 6s.
Dynamics. A Treatise on. By W. H. Besant, Sc.D., F.R.S. 7s. 6d.
Heat. An Elementary Treatise. By W. Garnett, M.A., D.C.L. 5th Edition, revised and enlarged. 4s. 6d.
Elementary Physics. Examples in. By W. Gallatly, M.A. 4s.
Hydromechanics. By W. H. Besant, Sc.D., F.R.S. 4th Edition. Part I. Hydrostatics. 5s.
Mathematical Examples. By J. M. Dyer, M.A., Eton College, and R. Prowde Smith, M.A., Cheltenham College. 6s.
Mechanics. Problems in Elementary. By W. Walton, M.A. 6s.

Educational Works.

CAMBRIDGE SCHOOL AND COLLEGE TEXT-BOOKS.
A Series of Elementary Treatises for the use of Students.

Arithmetic. By Rev. C. Elsee, M.A. Fcap. 8vo. 13th Edit. 3s. 6d.

—— By A. Wrigley, M.A. 3s. 6d.

—— A Progressive Course of Examples. With Answers. By J. Watson, M.A. 7th Edition, revised. By W. P. Goudie, B.A. 2s. 6d.

Algebra. By the Rev. C. Elsee, M.A. 7th Edit. 4s.

—— Progressive Course of Examples. By Rev. W. F. M'Michael, M.A., and R. Prowde Smith, M.A. 4th Edition. 3s. 6d. With Answers. 4s. 6d.

Plane Astronomy, An Introduction to. By P. T. Main, M.A. 5th Edition. 4s.

Conic Sections treated Geometrically. By W. H. Besant, Sc.D. 7th Edition. 4s. 6d. Solution to the Examples. 4s.

—— Enunciations and Figures Separately. 1s. 6d.

Statics, Elementary. By Rev. H. Goodwin, D.D. 2nd Edit. 3s.

Hydrostatics, Elementary. By W. H. Besant, Sc.D. 13th Edit. 4s.

Mensuration, An Elementary Treatise on. By B. T. Moore, M.A. 3s. 6d.

Newton's Principia, The First Three Sections of, with an Appendix; and the Ninth and Eleventh Sections. By J. H. Evans, M.A. 5th Edition, by P. T. Main, M.A. 4s.

Analytical Geometry for Schools. By T. G. Vyvyan. 5th Edit. 4s. 6d.

Greek Testament, Companion to the. By A. C. Barrett, M.A. 5th Edition, revised. Fcap. 8vo. 5s.

Book of Common Prayer, An Historical and Explanatory Treatise on the. By W. G. Humphry, B.D. 6th Edition. Fcap. 8vo. 2s. 6d.

Music, Text-book of. By Professor H. C. Banister. 14th Edition, revised. 5s.

—— Concise History of. By Rev. H. G. Bonavia Hunt, Mus. Doc. Dublin. 10th Edition revised. 3s. 6d.

ARITHMETIC AND ALGEBRA.
See also the two foregoing Series.

Arithmetic, Examination Papers in. Consisting of 140 papers, each containing 7 questions. 357 more difficult problems follow. A collection of recent Public Examination Papers are appended. By C. Pendlebury, M.A. 2s. 6d. Key, for Masters only, 5s.

Algebra, Examination Papers in. *Preparing.*

Graduated Exercises in Addition (Simple and Compound). By W. S. Beard, C. S. Dept. Rochester Mathematical School. 1s.
The Answers sent free to Masters only.
For Candidates for Commercial Certificates and Civil Service Exams.

BOOK-KEEPING.
Book-keeping Papers, set at various Public Examinations. Collected and Written by J. T. Medhurst, Lecturer on Book-keeping in the City of London College. 3s.

GEOMETRY AND EUCLID.

Euclid. Books I.-VI. and part of XI. and XII. A New Translation. By H. Deighton. Books I. and II. separately, 2s. (See p. 8.)
—— The Definitions of, with Explanations and Exercises, and an Appendix of Exercises on the First Book. By R. Webb, M.A. Crown 8vo. 1s. 6d.
—— Book I. With Notes and Exercises for the use of Preparatory Schools, &c. By Braithwaite Arnett, M.A. 8vo. 4s. 6d.
—— The First Two Books explained to Beginners. By C. P. Mason, B.A. 2nd Edition. Fcap. 8vo. 2s. 6d.
The Enunciations and Figures to Euclid's Elements. By Rev. J. Brasse, D.D. New Edition. Fcap. 8vo. 1s. Without the Figures, 6d.
Exercises on Euclid and in Modern Geometry. By J. McDowell, B.A. Crown 8vo. 3rd Edition revised. 6s.
Geometrical Conic Sections. By H. G. Willis, M.A. (See p. 8.)
Geometrical Conic Sections. By W. H. Besant, D.Sc. (See p. 9.)
Elementary Geometry of Conics. By C. Taylor, D.D. (See p. 8.)
An Introduction to Ancient and Modern Geometry of Conics. By C. Taylor, D.D., Master of St. John's Coll., Camb. 8vo. 15s.
Solutions of Geometrical Problems, proposed at St. John's College from 1830 to 1846. By T. Gaskin, M.A. 8vo. 12s.

TRIGONOMETRY.

Trigonometry, Introduction to Plane. By Rev. T. G. Vyvyan, Charterhouse. 3rd Edition. Cr. 8vo. 3s. 6d.
An Elementary Treatise on Mensuration. By B. T. Moore, M.A. 3s. 6d.
Trigonometry, Examination Papers in. By G. H. Ward, M.A., Assistant Master at St. Paul's School. Crown 8vo. 2s. 6d.

ANALYTICAL GEOMETRY AND DIFFERENTIAL CALCULUS.

An Introduction to Analytical Plane Geometry. By W. P. Turnbull, M.A. 8vo. 12s.
Problems on the Principles of Plane Co-ordinate Geometry. By W. Walton, M.A. 8vo. 16s.
Trilinear Co-ordinates, and Modern Analytical Geometry of Two Dimensions. By W. A. Whitworth, M.A. 8vo. 16s.
An Elementary Treatise on Solid Geometry. By W. S. Aldis, M.A. 4th Edition revised. Cr. 8vo. 6s.
Elliptic Functions, Elementary Treatise on. By A. Cayley, Sc.D. Professor of Pure Mathematics at Cambridge University. Demy 8vo. 15s.

MECHANICS & NATURAL PHILOSOPHY.

Statics, Elementary. By H. Goodwin, D.D. Fcap. 8vo. 2nd Edition. 3s.

Dynamics, A Treatise on Elementary. By W. Garnett, M.A., D.C.L. 5th Edition. Crown 8vo. 6s.

Dynamics. Rigid. By W. S. Aldis, M.A. 4s.

Dynamics. A Treatise on. By W. H. Besant, Sc.D., F.R.S. 7s. 6d.

Elementary Mechanics, Problems in. By W. Walton, M.A. New Edition. Crown 8vo. 6s.

Theoretical Mechanics, Problems in. By W. Walton, M.A. 3rd Edition. Demy 8vo. 16s.

Hydrostatics. By W. H. Besant, Sc.D. Fcap. 8vo. 13th Edition. 4s.

Hydromechanics, A Treatise on. By W. H. Besant, Sc.D., F.R.S. 8vo. 4th Edition, revised. Part I. Hydrostatics. 5s.

Hydrodynamics, A Treatise on. Vol. I. 10s. 6d.; Vol. II. 12s. 6d. A. B. Basset, M.A.

Optics, Geometrical. By W. S. Aldis, M.A. Crown 8vo. 3rd Edition. 4s.

Double Refraction, A Chapter on Fresnel's Theory of. By W. S. Aldis, M.A. 8vo. 2s.

Heat, An Elementary Treatise on. By W. Garnett, M.A., D.C.L. Crown 8vo. 5th Edition. 4s. 6d.

Elementary Physics. By W. Gallatly, M.A., Asst. Examr. at London University. 4s.

Newton's Principia, The First Three Sections of, with an Appendix; and the Ninth and Eleventh Sections. By J. H. Evans, M.A. 5th Edition. Edited by P. T. Main, M.A. 4s.

Astronomy, An Introduction to Plane. By P. T. Main, M.A. Fcap. 8vo. cloth. 5th Edition. 4s.

—— Practical and Spherical. By R. Main, M.A. 8vo. 14s.

Mathematical Examples. Pure and Mixed. By J. M. Dyer, M.A., and R. Prowde Smith, M.A. 6s.

Pure Mathematics and Natural Philosophy, A Compendium of Facts and Formulæ in. By G. R. Smalley. 2nd Edition, revised by J. McDowell, M.A. Fcap. 8vo. 3s. 6d.

Elementary Mathematical Formulæ. By the Rev. T. W. Openshaw, M.A. 1s. 6d.

Elementary Course of Mathematics. By H. Goodwin, D.D. 6th Edition. 8vo. 16s.

Problems and Examples, adapted to the 'Elementary Course of Mathematics.' 3rd Edition. 8vo. 5s.

Solutions of Goodwin's Collection of Problems and Examples. By W. W. Hutt, M.A. 3rd Edition, revised and enlarged. 8vo. 9s.

A Collection of Examples and Problems in Arithmetic, Algebra, Geometry, Logarithms, Trigonometry, Conic Sections, Mechanics, &c., with Answers. By Rev. A. Wrigley. 20th Thousand. 8s. 6d. Key. 10s. 6d.

Science Examination Papers. Part I. Inorganic Chemistry. By R. E. Steel, M.A., F.C.S., Bradford Grammar School. Crown 8vo. 2s. 6d.

TECHNOLOGICAL HANDBOOKS.

Edited by H. TRUEMAN WOOD, Secretary of the Society of Arts.

Dyeing and Tissue Printing. By W. Crookes, F.R.S. 5s.

Glass Manufacture. By Henry Chance, M.A.; H. J. Powell, B.A.; and H. G. Harris. 3s. 6d.

Cotton Spinning. By Richard Marsden, of Manchester. 3rd Edition, revised. 6s. 6d.

Chemistry of Coal-Tar Colours. By Prof. Benedikt, and Dr. Knecht of Bradford Technical College. 2nd Edition, enlarged. 6s. 6d.

Woollen and Worsted Cloth Manufacture. By Roberts Beaumont, Professor at Yorkshire College, Leeds. 7s. 6d.

Cotton Weaving. By R. Marsden. [*In the press.*

Colour in Woven Design. By Roberts Beaumont. [*In the press.*

Bookbinding. By Zaehnsdorf. [*Preparing.*

Others in preparation.

HISTORY, TOPOGRAPHY, &c.

Rome and the Campagna. By R. Burn, M.A. With 85 Engravings and 26 Maps and Plans. With Appendix. 4to. 21s.

Old Rome. A Handbook for Travellers. By R. Burn, M.A. With Maps and Plans. Demy 8vo. 5s.

Modern Europe. By Dr. T. H. Dyer. 2nd Edition, revised and continued. 5 vols. Demy 8vo. 2l. 12s. 6d.

The History of the Kings of Rome. By Dr. T. H. Dyer. 8vo. 5s.

The History of Pompeii: its Buildings and Antiquities. By T. H. Dyer. 3rd Edition, brought down to 1874. Post 8vo. 7s. 6d.

The City of Rome: its History and Monuments. 2nd Edition, revised by T. H. Dyer. 5s.

Ancient Athens: its History, Topography, and Remains. By T. H. Dyer. Super-royal 8vo. Cloth. 7s. 6d.

The Decline of the Roman Republic. By G. Long. 5 vols. 8vo. 5s. each.

Historical Maps of England. By C. H. Pearson. Folio. 3rd Edition revised. 31s. 6d.

History of England, 1800–46. By Harriet Martineau, with new and copious Index. 5 vols. 3s. 6d. each.

A Practical Synopsis of English History. By A. Bowes. 9th Edition, revised. 8vo. 1s.

Lives of the Queens of England. By A. Strickland. Library Edition, 8 vols. 7s. 6d. each. Cheaper Edition, 6 vols. 5s. each. Abridged Edition, 1 vol. 6s. 6d. Mary Queen of Scots, 2 vols. 5s. each. Tudor and Stuart Princesses, 5s.

Educational Works.

Eginhard's Life of Karl the Great (Charlemagne). Translated, with Notes, by W. Glaister, M.A., B.C.L. Crown 8vo. 4s. 6d.

The Elements of General History. By Prof. Tytler. New Edition, brought down to 1874. Small Post 8vo. 3s. 6d.

History and Geography Examination Papers. Compiled by C. H. Spence, M.A., Clifton College. Crown 8vo. 2s. 6d.

PHILOLOGY.

WEBSTER'S DICTIONARY OF THE ENGLISH LANGUAGE. With Dr. Mahn's Etymology. 1 vol. 1628 pages, 3000 Illustrations. 21s.; half calf, 30s.; calf or half russia, 31s. 6d.; russia, 2l. With Appendices and 70 additional pages of Illustrations, 1919 pages, 31s. 6d.; half calf, 2l.; calf or half russia, 2l. 2s.; russia, 2l. 10s.
'THE BEST PRACTICAL ENGLISH DICTIONARY EXTANT.'—*Quarterly Review*, 1873.
Prospectuses, with specimen pages, post free on application.

Richardson's Philological Dictionary of the English Language. Combining Explanation with Etymology, and copiously illustrated by Quotations from the best Authorities. With a Supplement. 2 vols. 4to. 4l. 14s. 6d. Supplement separately. 4to. 12s.

Brief History of the English Language. By Prof. James Hadley, LL.D., Yale College. Fcap. 8vo. 1s.

The Elements of the English Language. By E. Adams, Ph.D. 21st Edition. Post 8vo. 4s. 6d.

Philological Essays. By T. H. Key, M.A., F.R.S. 8vo. 10s. 6d.

Synonyms and Antonyms of the English Language. By Archdeacon Smith. 2nd Edition. Post 8vo. 5s.

Synonyms Discriminated. By Archdeacon Smith. Demy 8vo. 2nd Edition revised. 14s.

Bible English. Chapters on Words and Phrases in the Bible and Prayer Book. By Rev. T. L. O. Davies. 2nd Edition revised, in the press.

The Queen's English. A Manual of Idiom and Usage. By the late Dean Alford. 6th Edition. Fcap. 8vo. 1s. sewed. 1s. 6d. cloth.

A History of English Rhythms. By Edwin Guest, M.A., D.C.L. LL.D. New Edition, by Professor W. W. Skeat. Demy 8vo. 18s.

Elements of Comparative Grammar and Philology. For Use in Schools. By A. C. Price, M.A., Assistant Master at Leeds Grammar School. Crown 8vo. 2s. 6d.

Questions for Examination in English Literature. By Prof. W. W. Skeat. 2nd Edition, revised. 2s. 6d.

A Syriac Grammar. By G. Phillips, D.D. 3rd Edition, enlarged. 8vo. 7s. 6d.

DIVINITY, MORAL PHILOSOPHY, &c.

BY THE REV. F. H. SCRIVENER, A.M., LL.D., D.C.L.

Novum Testamentum Græce. Editio major. Being an enlarged Edition, containing the Readings of Westcott and Hort, and those adopted by the Revisers, &c. 7s. 6d. *For other Editions see page 3.*

A Plain Introduction to the Criticism of the New Testament. With Forty Facsimiles from Ancient Manuscripts. 3rd Edition. 8vo. 18s.

Six Lectures on the Text of the New Testament. For English Readers. Crown 8vo. 6s.

Codex Bezæ Cantabrigiensis. 4to. 10s. 6d.

The New Testament for English Readers. By the late H. Alford, D.D. Vol. I. Part I. 3rd Edit. 12s. Vol. I. Part II. 2nd Edit. 10s. 6d. Vol. II. Part I. 2nd Edit. 16s. Vol. II. Part II. 2nd Edit. 16s.

The Greek Testament. By the late H. Alford, D.D. Vol. I. 7th Edit. 1l. 8s. Vol. II. 8th Edit. 1l. 4s. Vol. III. 10th Edit. 18s. Vol. IV. Part I. 5th Edit. 18s. Vol. IV. Part II. 10th Edit. 14s. Vol. IV. 1l. 12s.

Companion to the Greek Testament. By A. C. Barrett, M.A. 5th Edition, revised. Fcap. 8vo. 5s.

Guide to the Textual Criticism of the New Testament. By Rev. E. Miller, M.A. Crown 8vo. 4s.

The Book of Psalms. A New Translation, with Introductions, &c. By the Very Rev. J. J. Stewart Perowne, D.D. 8vo. Vol. I. 6th Edition, 18s. Vol. II. 6th Edit. 16s.

—— Abridged for Schools. 6th Edition. Crown 8vo. 10s. 6d.

History of the Articles of Religion. By C. H. Hardwick. 3rd Edition. Post 8vo. 5s.

History of the Creeds. By J. R. Lumby, DD. 3rd Edition. Crown 8vo. 7s. 6d.

Pearson on the Creed. Carefully printed from an early edition. With Analysis and Index by E. Walford, M.A. Post 8vo. 5s.

Liturgies and Offices of the Church, for the Use of English Readers, in Illustration of the Book of Common Prayer. By the Rev. Edward Burbidge, M.A. Crown 8vo. 9s.

An Historical and Explanatory Treatise on the Book of Common Prayer. By Rev. W. G. Humphry, B.D. 6th Edition, enlarged. Small Post 8vo. 2s. 6d.; Cheap Edition, 1s.

A Commentary on the Gospels, Epistles, and Acts of the Apostles. By Rev. W. Denton, A.M. New Edition. 7 vols. 8vo. 9s. each.

Notes on the Catechism. By Rt. Rev. Bishop Barry. 8th Edit. Fcap. 2s.

The Winton Church Catechist. Questions and Answers on the Teaching of the Church Catechism. By the late Rev. J. S. B. Monsell, LL.D. 4th Edition. Cloth, 3s.; or in Four Parts, sewed.

The Church Teacher's Manual of Christian Instruction. By Rev. M. F. Sadler. 38th Thousand. 2s. 6d.

Educational Works. 15

FOREIGN CLASSICS.

A Series for use in Schools, with English Notes, grammatical and explanatory, and renderings of difficult idiomatic expressions.
Fcap. 8vo.

Schiller's Wallenstein. By Dr. A. Buchheim. 5th Edit. 5s.
Or the Lager and Piccolomini, 2s. 6d. Wallenstein's Tod, 2s. 6d.
—— **Maid of Orleans.** By Dr. W. Wagner. 2nd Edit. 1s. 6d.
—— **Maria Stuart.** By V. Kastner. 2nd Edition. 1s. 6d.
Goethe's Hermann and Dorothea. By E. Bell, M.A., and E. Wölfel. 1s. 6d.
German Ballads, from Uhland, Goethe, and Schiller. By C. L. Bielefeld. 3rd Edition. 1s. 6d.
Charles XII., par Voltaire. By L. Direy. 7th Edition. 1s. 6d.
Aventures de Télémaque, par Fénélon. By C. J. Delille. 4th Edition. 2s. 6d.
Select Fables of La Fontaine. By F. E. A. Gasc. 18th Edit. 1s. 6d.
Picciola, by X. B. Saintine. By Dr. Dubuc. 15th Thousand. 1s. 6d.
Lamartine's Le Tailleur de Pierres de Saint-Point. By J. Boïelle, 4th Thousand. Fcap. 8vo. 1s. 6d.
Italian Primer. By Rev. A. C. Clapin, M.A. Fcap. 8vo. 1s.

FRENCH CLASS-BOOKS.

French Grammar for Public Schools. By Rev. A. C. Clapin, M.A. Fcap. 8vo. 12th Edition, revised. 2s. 6d.
French Primer. By Rev. A. C. Clapin, M.A. Fcap. 8vo. 8th Ed. 1s.
Primer of French Philology. By Rev. A. C. Clapin. Fcap. 8vo. 4th Edit. 1s.
Le Nouveau Trésor; or, French Student's Companion. By M. E. S. 18th Edition. Fcap. 8vo. 1s. 6d.
French Examination Papers in Miscellaneous Grammar and Idioms. Compiled by A. M. M. Stedman, M.A. 4th Edition. Crown 8vo. 2s. 6d.
Key to the above. By G. A. Schrumpf, Univ. of France. Crown 8vo. 5s. (For Teachers or Private Students only.)
Manual of French Prosody. By Arthur Gosset, M.A. Crown 8vo. 3s.
Lexicon of Conversational French. By A. Holloway. 3rd Edition. Crown 8vo. 4s.

PROF. A. BARRÈRE'S FRENCH COURSE.

Elements of French Grammar and First Steps in Idiom. Crown 8vo. 2s.
Precis of Comparative French Grammar. 2nd Edition. Crown 8vo. 3s. 6d.
Junior Graduated French Course. Crown 8vo. 1s. 6d.

F. E. A. GASC'S FRENCH COURSE.

First French Book. Fcap. 8vo. 106th Thousand. 1*s.*
Second French Book. 47th Thousand. Fcap. 8vo. 1*s.* 6*d.*
Key to First and Second French Books. 5th Edit. Fcp. 8vo. 3*s.* 6*d.*
French Fables for Beginners, in Prose, with Index. 16th Thousand. 12mo. 1*s.* 6*d.*
Select Fables of La Fontaine. 18th Thousand. Fcap. 8vo. 1*s.* 6*d.*
Histoires Amusantes et Instructives. With Notes. 16th Thousand. Fcap. 8vo. 2*s.*
Practical Guide to Modern French Conversation. 17th Thousand. Fcap. 8vo. 1*s.* 6*d.*
French Poetry for the Young. With Notes. 5th Ed. Fcp. 8vo. 8*s.*
Materials for French Prose Composition; or, Selections from the best English Prose Writers. 19th Thous. Fcap. 8vo. 3*s.* Key, 6*s.*
Prosateurs Contemporains. With Notes. 10th Edition, revised. 12mo. 3*s.* 6*d.*
Le Petit Compagnon; a French Talk-Book for Little Children. 12th Thousand. 16mo. 1*s* 6*d.*
An Improved Modern Pocket Dictionary of the French and English Languages. 45th Thousand. 16mo. 2*s.* 6*d.*
Modern French-English and English-French Dictionary. 4th Edition, revised, with new supplements. 10*s.* 6*d.*
The A B C Tourist's French Interpreter of all Immediate Wants. By F. E. A. Gasc. 1*s.*

MODERN FRENCH AUTHORS.

Edited, with Introductions and Notes, by JAMES BOÏELLE, Senior French Master at Dulwich College.

Daudet's La Belle Nivernaise. 2*s.* 6*d.* *For Beginners.*
Hugo's Bug Jargal. 3*s.* *For Advanced Students.*

GOMBERT'S FRENCH DRAMA.

Being a Selection of the best Tragedies and Comedies of Molière, Racine, Corneille, and Voltaire. With Arguments and Notes by A. Gombert. New Edition, revised by F. E. A. Gasc. Fcap. 8vo. 1*s.* each; sewed, 6*d.*

CONTENTS.

MOLIERE:—Le Misanthrope. L'Avare. Le Bourgeois Gentilhomme. Le Tartuffe. Le Malade Imaginaire. Les Femmes Savantes. Les Fourberies de Scapin. Les Précienses Ridicules. L'Ecole des Femmes. L'Ecole des Maris. Le Médecin malgré Lui.
RACINE:—Phèdre. Esther. Athalie. Iphigénie. Les Plaideurs. La Thébaïde; ou, Les Frères Ennemis. Andromaque. Britannicus.
P. CORNEILLE:—Le Cid. Horace. Cinna. Polyeucte.
VOLTAIRE:—Zaïre.

GERMAN CLASS-BOOKS.

Materials for German Prose Composition. By Dr. Buchheim. 12th Edition, thoroughly revised. Fcap. 4*s.* 6*d.* Key, Parts I. and II., 3*s.* Parts III. and IV., 4*s.*

German Conversation Grammar. By I. Sydow. 2nd Edition. Book I. Etymology. 2*s.* 6*d.* Book II. Syntax. 1*s.* 6*d.*

Wortfolge, or Rules and Exercises on the Order of Words in
German Sentences. By Dr. F. Stock. 1s. 6d.
A German Grammar for Public Schools. By the Rev. A. C.
Clapin and F. Holl Müller. 5th Edition. Fcap. 2s. 6d.
A German Primer, with Exercises. By Rev. A. C. Clapin. 1s
Kotzebue's Der Gefangene. With Notes by Dr. W. Stromberg. 1s.
German Examination Papers in Grammar and Idiom. By
R. J. Morich. 2s. 6d. Key for Tutors only, 5s.
By FRZ. LANGE, Ph.D., Professor R.M.A., Woolwich, Examiner
in German to the Coll. of Preceptors, and also at the
Victoria University, Manchester.
A Concise German Grammar. In Three Parts. Part I. Elementary. 2s. Part II. Intermediate. 1s. 6d. Part III. Advanced, 3s. 6d.
German Examination Course. Elementary, 2s. Intermediate, 2s.
Advanced, 1s. 6d.
German Reader. Elementary. 1s. 6d. Advanced (*in the press*).

MODERN GERMAN SCHOOL CLASSICS.
Small Crown 8vo.

Hey's Fabeln Für Kinder. Edited by Prof. F. Lange, Ph.D. 1s. 6d.
Benedix's Dr. Wespe. Edited by F. Lange, Ph.D. 2s. 6d.
Hoffman's Meister Martin, der Küfner. By Prof. F. Lange, Ph.D.
1s. 6d.
Heyse's Hans Lange. By A. A. Macdonell, M.A., Ph.D. 2s.
Auerbach's Auf Wache, and Roquette's Der Gefrorene Kuss.
By A. A. Macdonell, M.A. 2s.
Moser's Der Bibliothekar. By Prof. F. Lange, Ph.D. 2s.
Ebers' Eine Frage. By F. Storr, B.A. 2s.
Freytag's Die Journalisten. By Prof. F. Lange, Ph.D. 2s. 6d.
Gutzkow's Zopf und Schwert. By Prof. F. Lange, Ph.D. 2s.
German Epic Tales. Edited by Kar Neuhaus, Ph.D. 2s. 6d.

ENGLISH CLASS-BOOKS.
Comparative Grammar and Philology. By A. C. Price, M.A.,
Assistant Master at Leeds Grammar School. 2s. 6d.
The Elements of the English Language. By E. Adams, Ph.D.
22nd Edition. Post 8vo. 4s. 6d.
The Rudiments of English Grammar and Analysis. By
E. Adams, Ph.D. 17th Thousand. Fcap. 8vo. 1s.
A Concise System of Parsing. By L. E. Adams, B.A. 1s. 6d.
General Knowledge Examination Papers. Compiled by
A. M. M. Stedman, M.A. 2s. 6d.
Examples for Grammatical Analysis (Verse and Prose). Selected, &c., by F. Edwards. New edition. Cloth, 1s.
Notes on Shakespeare's Plays. By T. Duff Barnett, B.A.
MIDSUMMER NIGHT'S DREAM, 1s.; JULIUS CÆSAR, 1s.; HENRY V., 1s.;
TEMPEST, 1s.; MACBETH, 1s.; MERCHANT OF VENICE, 1s.; HAMLET, 1s.

By C. P. MASON, Fellow of Univ. Coll. London.
First Notions of Grammar for Young Learners. Fcap. 8vo. 47th Thousand. Cloth. 9d.
First Steps in English Grammar for Junior Classes. Demy 18mo. 46th Thousand. 1s.
Outlines of English Grammar for the Use of Junior Classes. 71st to 76th Thousand. Crown 8vo. 2s.
English Grammar, including the Principles of Grammatical Analysis. 31st Edition. 125th to 130th Thousand. Crown 8vo. 3s. 6d.
Practice and Help in the Analysis of Sentences. 2s.
A Shorter English Grammar, with copious Exercises. 34th to 38th Thousand. Crown 8vo. 3s. 6d.
English Grammar Practice, being the Exercises separately. 1s.
Code Standard Grammars. Parts I. and II., 2d. each. Parts III., IV., and V., 3d. each.

Notes of Lessons, their Preparation, &c. By José Rickard, Park Lane Board School, Leeds, and A. H. Taylor, Rodley Board School, Leeds. 2nd Edition. Crown 8vo. 2s. 6d.
A Syllabic System of Teaching to Read, combining the advantages of the 'Phonic' and the 'Look-and-Say' Systems. Crown 8vo. 1s.
Practical Hints on Teaching. By Rev. J. Menet, M.A. 6th Edit. revised. Crown 8vo. paper, 2s.
How to Earn the Merit Grant. A Manual of School Management. By H. Major, B.A., B.Sc. Part I. (3rd Edit.) Infant School, 3s. Part II. (2nd Edit. revised), 4s. Complete, 6s.
Test Lessons in Dictation. 4th Edition. Paper cover, 1s. 6d.
The Botanist's Pocket-Book. With a copious Index. By W. R. Hayward. 6th Edition, revised. Crown 8vo. cloth limp. 4s. 6d.
Experimental Chemistry, founded on the Work of Dr. Stöckhardt. By C. W. Heaton. Post 8vo. 5s.
Lectures on Musical Analysis. Sonata-form, Fugue, &c. By Prof. H. C. Banister. 2nd Edition, revised. 7s. 6d.
Helps' Course of Poetry, for Schools. A New Selection from the English Poets, carefully compiled and adapted to the several standards by E. A. Helps, one of H.M. Inspectors of Schools.
    Book I. Infants and Standards I. and II. 134 pp. small 8vo. 9d.
    Book II. Standards III. and IV. 224 pp. crown 8vo. 1s. 6d.
    Book III. Standards V., VI., and VII. 352 pp. post 8vo. 2s.
      Or in PARTS. Infants, 2d.; Standard I., 2d.; Standard II., 2d. Standard III., 4d.
Picture School-Books. In Simple Language, with numerous Illustrations. Royal 16mo.
    The Infant's Primer. 3d.—School Primer. 6d.—School Reader. By J. Tilleard. 1s.—Poetry Book for Schools. 1s.—The Life of Joseph. 1s.—The Scripture Parables. By the Rev. J. E. Clarke. 1s.—The Scripture Miracles. By the Rev. J. E. Clarke. 1s.—The New Testament History. By the Rev. J. G. Wood, M.A. 1s.—The Old Testament History. By the Rev. J. G. Wood, M.A. 1s.—The Life of Martin Luther. By Sarah Crompton. 1s.

Educational Works. 1

BOOKS FOR YOUNG READERS.
A Series of Reading Books designed to facilitate the acquisition of the power of Reading by very young Children. In 11 *vols. limp cloth,* 6d. *each.*

Those with an asterisk have a Frontispiece or other Illustrations.

*The Old Boathouse. Bell and Fan; or, A Cold Dip.
*Tot and the Cat. A Bit of Cake. The Jay. The Black Hen's Nest. Tom and Ned. Mrs. Bee.
*The Cat and the Hen. Sam and his Dog Redleg. Bob and Tom Lee. A Wreck.
*The New-born Lamb. The Rosewood Box. Poor Fan. Sheep Dog.

} *Suitable for Infants.*

*The Two Parrots. A Tale of the Jubilee. By M. E. Wintle. 9 Illustrations.
*The Story of Three Monkeys.
*Story of a Cat. Told by Herself.
The Blind Boy. The Mute Girl. A New Tale of Babes in a Wood.
The Dey and the Knight. The New Bank Note. The Royal Visit. A King's Walk on a Winter's Day.
*Queen Bee and Busy Bee.
*Gull's Crag.
*A First Book of Geography. By the Rev. C. A. Johns. Illustrated. Double size, 1s.

} *Suitable for Standards I. & II.*

Syllabic Spelling. By C. Barton. In Two Parts. Infants, 3d. Standard I., 3d.

GE GRAPHICAL SERIES. By M. J. BARRINGTON WARD, M.A. *With Illustrations.*

Th Map and the Compass. A Reading-Book of Geography. For Standard I. New Edition, revised. 8d. cloth.
The Round World. A Reading-Book of Geography. For Standard II. 10d.
About England. A Reading Book of Geography for Standard III. [*In the press.*
The Child's Geography. For the Use of Schools and for Home Tuition. 6d.
The Child's Geography of England. With Introductory Exercises on the British Isles and Empire, with Questions. 2s. 6d. Without Questions, 2s.

Geography Examination Papers. (See History and Geography Papers, p. 12.)

BELL'S READING-BOOKS.
FOR SCHOOLS AND PAROCHIAL LIBRARIES.
Now Ready. Post 8vo. Strongly bound in cloth, 1s. each.

*Life of Columbus.
*Grimm's German Tales. (Selected.)
*Andersen's Danish Tales. Illustrated. (Selected.)
Great Englishmen. Short Lives for Young Children.
Great Englishwomen. Short Lives of.
Great Scotsmen. Short Lives of.
*Masterman Ready. By Capt. Marryat. Illus. (Abgd.)
*Poor Jack. By Capt. Marryat, R.N. (Abridged.)

} Suitable for Standards III. & IV.

*Scott's Talisman. (Abridged.)
*Friends in Fur and Feathers. By Gwynfryn.
*Poor Jack. By Captain Marryat, R.N. Abgd.
Parables from Nature. (Selected.) By Mrs. Gatty.
Lamb's Tales from Shakespeare. (Selected.)
Edgeworth's Tales. (A Selection.)
*Gulliver's Travels. (Abridged.)
*Robinson Crusoe. Illustrated.
*Arabian Nights. (A Selection Rewritten.)

} Standards IV. & V.

*Dickens's Little Nell. Abridged from the 'The Old Curiosity Shop.'
*The Vicar of Wakefield.
*Settlers in Canada. By Capt. Marryat. (Abridged.)
Marie: Glimpses of Life in France. By A. R. Ellis.
Poetry for Boys. Selected by D. Munro.
*Southey's Life of Nelson. (Abridged.)
*Life of the Duke of Wellington, with Maps and Plans.
*Sir Roger de Coverley and other Essays from the Spectator.
Tales of the Coast. By J. Runciman.

} Standards V. VI. & VII.

* *These Volumes are Illustrated.*

Uniform with the Series, in limp cloth, 6d. each.
Shakespeare's Plays. Kemble's Reading Edition. With Explanatory Notes for School Use.
JULIUS CÆSAR. THE MERCHANT OF VENICE. KING JOHN.
HENRY THE FIFTH. MACBETH. AS YOU LIKE IT.

London: **GEORGE BELL & SONS**, York Str et, Covent Garden.

www.ingramcontent.com/pod-product-compliance
Lightning Source LLC
Chambersburg PA
CBHW030322240426
43673CB00040B/1245